SAFE TO CHEW

— An Anthology —

©2015 Wicwas Press, LLC

Editors and Contest Judges Lawrence Connor, Randy Kim, Robert Muir.

Contributing Editor and Co-Judge Cindy Bee.

Special Thanks to Steve Repasky.

Introduction by Robert Muir, partner at Wicwas Press, LLC.

Cover Design by Wicwas Press, LLC.

Cover Photo by Lawrence J. Connor.

Wicwas Press, LLC • Kalamazoo, Michigan • www.wicwas.com

Printed in U.S.A.

First printing

ISBN 978-1-878075-45-1 PERFECT BOUND

ISBN 978-1-878075-46-8 EPUB

TABLE OF CONTENTS

INTRODUCTION   1

KINSHIP   3

It Takes Time to Build a Home *by Benjamin Hance*   4
My Father's Bees *by Peta-Ann Forbes*   8
Our Beekeeper Laura *by William Hawkins*   10
"Just an Assistant" *by Helen Ackermann*   12
Our Grandfather, the Beekeeper *by Laquavia Foster*   14
An Unexpected Career *by Jane Resides*   17
The Bee Boxes *by Michelle Purvis*   19
Hive Mentality *by Catherine Blakely*   23

WAXING   27

Bastion of the Bees *by Rebekah Hoeft*   28
The Melt of Honey *by Bill Pendergraft*   29
Through the Window *by Zosia Botto*   30
Among the Clover *by Carrie Montgomery*   33
Truth *by Vicki Wilke*   35
I Don't Think That Much *by Nathan Cox*   36
Bee-Whiz *by Jordan Link*   39
The Idea of Bees *by Vicki Wilke*   42
Bee Attitudes *by Mary Langer Thompson*   44
The Stadium of Bees *by Anna Citrino*   45

NEW WAX   47

A Honey of a Conversation *by Billie Holladay Skelley*   48
Love Blooms *by Angeline Haen*   52
Of Bee *by J.P.A.*   53
A Bee in Her Bonnet *by Jill Crainshaw*   54
The Taste of Happiness *by Lauren Pichard*   56
Spring Fever *by Robbi Nester*   57
Honey Haiku *by Angeline Haen*   58

BITTER HONEY 59

Smoke *by Tabitha Peyton Wood* — 60
Song of the Swarm *by Elizabeth Christie* — 65
Keeping Watch *by Dan Stout* — 68
The Radish Field *by Tarah Walsh* — 72
The Honey of Oil *by Rachel Channell* — 74
In Golden Slumber *by Liza Mattison* — 75
The Land of Milk and Honey *by Alixandra Nicole Prybyla* — 80

WITH THE SEASON 83

The Raven and the Bee *by Sarah Herdan* — 84
Hospitaliano *by Daniel Talamantes* — 87
Lonnie's Front Porch *by Deborah Darling* — 92
Letters After Achilles *by Stefanie Brook Trout* — 93
Patron Saint of Bees *by K.E. Kuebler* — 98
Colony Collapse Disorder *by Amy Newday* — 102
Brood Chamber *by Daniel Talamantes* — 103
Building Defenses *by Erica Eastick* — 106
Winter Hive *by Andrea Dickens* — 111

DEFENSIVE BEHAVIOR 112

The Bee Bully *by Inga Harris* — 113
Beekeeping with a Honeybee Allergy *by Jennifer Ford* — 114
Strange Bees *by Karen J. Bryant* — 117
Apiary in May *by Payton Sullivan* — 121
Diablo Canyon | Alice Many Ears of Corn *by Gary Ives* — 124

THE BUZZ 127

Tale of the Dragon Bee *by Randy Ames* — 128
From a Bear's Perspective *by Josh Wachtenheim* — 132
Regibald Sees the Eggs *by William Blomstedt* — 133
Waggle Dance *by Andrea Dickens* — 137
The Bee Thief *by MFC Feeley* — 138
Fred the Bear *by Roman Carrier* — 142
King Bee *by Lela Marie De La Garza* — 144

ANOTHER PERSPECTIVE 146

Honeybee's Journey *by Abigail Miller* 147
The Tale of the Sting *by Tibor Csincsa* 149
The Wasp and the Bee *by Tibor Csincsa* 152
Honeybee *by Charles Brunn* 156
A Sickness in the Countryside *by Sam Bassett* 157
Learning From the Bees *by Tulip Chowdhury* 161

APPLIED BEEKEEPING 164

Sharing a Cell *by Dr. Christian W. TwoRivers* 165
Commercial Keeping *by William Blomstedt* 168
To Course A Bee *by William E. Jones* 172
In Service to the Queen *by Joan Riise* 173
Experts *by Deborah Darling* 177
SSSSURPRISE! *by Phill Remick* 178
Taste of Chicago *by Jason M. Shimotake* 181
Lessons Learned *by Helen Ackermann* 185
The Mindfulness of Beekeeping *by Jennifer Ford* 187

PASSING THE SMOKER 189

Only the Bees Know *by Greg Leatherman* 190
Brave New Worlds *by Linda Butler* 191
Blessing of the Bees *by Hannah Valentine* 194
Adeline's Honey Harvest *by Rachael Gaude* 197
The Waggle Dance *by Laurie Wallmark* 201
Robin's Bees *by Therese Calegari* 202
The Beginning *by Katharine Atwood* 207

CONTRIBUTORS 209

INTRODUCTION

The success of this collection will be measured by whether or not you find yourself smiling with a tilt of your head and musing to yourself, "I never would have thought of that." Certainly, this is a new and ambitious undertaking for Wicwas Press. Our focus and notoriety in recent years has been on providing informative, quality books on the subject the industrious, beloved, mysterious honey bee and the art of its keeping. We've met with thousands of beekeepers, authors, scientists and enthusiasts, all of whom have been drawn together by a spark of curiosity and inspiration—and inspiration takes many forms. Within these pages, we have attempted to capture and cultivate that inspiration in a series of essays, poems and stories around the theme of beekeeping.

Our contributors are not necessarily writers or beekeepers, though many are or will soon find themselves both. We invited their participation in the first ever Wicwas Press anthology with the belief that "every beekeeper has a story to tell" and we were not disappointed. With over 200 original submissions to the effort by people from all walks of life, we are humbled by the response. We extend our thanks to everyone who participated, perhaps especially to those bold individuals whose works did not make the final cut. Every submission brought with it unique elements and, ultimately, I hope that we have gathered the most outstanding representatives of these diverse expressions.

These collected works present themes spanning the breadth of all beekeepers' experiences and imagination, including stories that explore beekeeping through the generations, the joyful innocence of natural curiosity and flights of fancy, as well as some of the darkness, loss, hardships and difficult lessons that accompany our ancient obsession.

Among the pieces to follow, you will find the winners of the contest that enticed so many of you to participate. It was the privilege of the Wicwas Press staff and our guest judge, Cindy Bee, to offer awards of excellence to the following pieces:

First: *It Takes Time to Build a Home* by Benjamin Hance

Second: *Smoke* by Tabitha Peyton Wood

Third: *Sharing a Cell* by Dr. Christian W. TwoRivers

We hope that you enjoyed these pieces as much as we did and, if you find this collection sweet to the taste, you have yourselves to thank.

May this collection serve as a tribute to the ingenuity and thoughtfulness of a world that, from time to time, remembers fondly its place in the grand design, and how big a part the littlest of things can play.

—Robert G. Muir

KINSHIP

It Takes Time to Build a Home

Benjamin Hance

I thought at the time that the story of my life must have been one of the great tragedies of the ages. Filled with the certainty of youth and all the wisdom and worldly knowledge of my twelve years of life, I just *knew* that nobody had it harder than I did. Looking back, of course, I now see how absurd that notion was. My life was no great tale of woe, no misfortune of Shakespearean caliber. I was merely a young man who grew up poor. My parents had neither space for me in their apartment nor the money to feed an ever-hungry boy on the threshold of his teen years. And so, in what I thought must have been the greatest injustice ever committed, I was sent to live with my grandparents in the country.

To refer to the section of rural southern Maryland that my grandparents called home as the middle of nowhere would be a compliment. Such a label might actually give the place too much credit as it would imply that the area was the *middle* of something. Wedged between a swamp and acres of wheat and tobacco, my grandfather's stone farmhouse sat back from a small dirt road. The small dirt road connected to a large dirt road which ran for forty miles before any asphalt appeared. When the "bus" from Baltimore that was, in truth, a station wagon dropped me off that summer day in the town of Well's Cove, my grandfather picked me up in an honest-to-God horse cart.

My grandfather had never learned to drive a car. He had never been outside of the state of Maryland. He had never finished school. And it seemed to me that he had never regretted any of this. Not that I knew much about his feelings at the time; I had met him exactly twice before I moved into his house, and he was not a talkative man. In spite of all of the things he had never done, my grandfather was a bit of a redneck renaissance man. When fish were plentiful, he was a waterman; when they were not, he was a farmer. He hunted, trapped, or raised the food he ate, and everything on his property was built with his own hands. This included every building from the house itself to the beehives he tended on the edge of his fields. The bees were originally a concession to my grandmother who refused to start her morning without a honey-sweetened cup of tea, but he soon found that a few jars of honey could make ends meet during the hardest of times when both the land and sea failed to produce.

I did not adjust well to moving and soon found myself in a miserable state. Natives of southern Maryland had a long established tradition of wariness towards outsiders even if they claimed kinship to a long-time resident. Making friends was not easy, and the sting of rejection and loneliness did not fade quickly. My parents visited infrequently because of the distance and cost, and my grandparents, however welcoming, were still strangers to me. I was a homesick and moody child, and I stayed in this condition for several weeks before my grandfather—no doubt at my grandmother's command—took action to cheer me up.

My grandfather's plan to raise my spirits, however, was decidedly self-serving. He determined that what I needed to improve my mood was a task to fill my time and give me a sense of purpose and, eventually, accomplishment. And there was *always* work to be done around the farm. So, around mid-summer, he walked me out to his workshop and showed me what my project would be.

His hive numbers were strong, and the time had come to split the hive. But for that he needed new frames, and it was my job to construct them. He gave me a full array of second-hand carpenter's tools, a pile of seasoned cypress wood, a set of frame samples made from white pine and a clap on the shoulder as he told me to get to work.

My grandfather would sit with me for an hour each day to check my work and explain to me exactly what it was I was doing. The advantages of Langstroth frames over top bars were extolled. He told me of the difference between the deep brood frames and the shallow frames for the honey supers. I learned the purpose of a queen excluder and why he thought cypress the best wood for a hive.

But I didn't care for his lessons. The work was hard and frustrating. It required a skill and precision with woodworking that I did not have, and each day was a new reminder of how inadequate I was to the task I had been assigned. Eventually, in the second week of my work building frames, I snapped. I attempted to assemble a frame only to find I had incorrectly sized the bottom bar, and the frame was more trapezoidal than rectangular. My frustration boiled over, and I began to smash the frame against the wall. My loneliness, inability to make friends and anger at myself, others and the world all combined and surged out of me in a guttural yell as I utterly destroyed that unlucky Langstroth frame.

It was only after several seconds that I noticed my grandfather standing in the doorway, arms crossed over his chest and brow furrowed low over his eyes. I was ashamed at myself for letting him see my tantrum, and that shame pooled inside me with the rest of my emotions. My last resistance failed, and I began to wordlessly sob as I dropped the splintered remains of the frame to the floor.

My grandfather crossed the workshop silently and stood by my side, nudging the broken pieces with the toe of his boot. He put his arm around my shoulders and let it stay there as I cried inconsolably. After several long minutes, when my sobs had subsided to a manageable level, he spoke to me softly, thankfully without looking at my tear-streaked face and leaky nose.

"It takes time to build a home."

He gently squeezed my shoulders and handed me his handkerchief before walking quietly back out to work in the fields.

It did take time, but I eventually built a home for the new hive. And it took an even longer time, but I eventually made a home for myself in Well's Cove. I did make friends, however slowly, and my life became a bit brighter day by day. And one day, without noticing, I stopped thinking of my grandparent's house as a place I had been sent and began to think of it as the place I lived.

My grandfather passed away many years later, long after I had grown and moved away. Absorbed in my own life and work, I had not seen him in nearly half a decade. He left me the entire farm. When I returned to Well's Cove for his funeral, I saw that the barns and workshop were ramshackle, and the hives were empty. Neighbors told me that he had been too frail in his final years to handle the upkeep but too proud to accept any help. I was not surprised.

I'm not sure whether it was a sense of nostalgia that convinced me to move back to Well's Cove or a sense of familial duty. Either way, I am happy here. I have once again built a home, one I share with my wife and our four year old daughter. Times have changed; the roads have asphalt and stoplights, and the other farms are increasingly broken up into new subdivisions. Even the cypress trees are all but vanished. But I'm doing my best to keep the farm the way I think my grandfather would want it and the way I still want it. I rebuilt the barn, and the workshop, and, of course, the hives.

I'm going to split a hive soon, and so I am building frames. I know I could buy them, but it just wouldn't be the same. My daughter likes to watch me in the workshop, sitting on a stool as I shape the wood. With all the patience that one might expect of a four year old, she asks practically every day how long until the new "bee house" is ready. Each time she asks, all I do is smile and respond in the only way I feel I can.

"It takes time to build a home."

My Father's Bees
Peta-Ann Forbes

My father pulled up in an old white pickup truck he affectionately referred to as his van. I had not seen the man in four years. I got out of my cousin's car where I was anxiously awaiting his arrival, said my goodbyes and greeted my father. "Hi daddy," I said happily as I threw my arms around his neck. I walked around the back of the van and noticed several white boxes of something. I asked him what was in them. He said with deep bass in his voice, "My Bees." I later learned that my father had acquired several new bees from a small nest he found inside a wall of his mother-in-law's house.

My father was a very rough looking, scruffy bearded man whose skin was heavily weathered by the sun. He was a man of few words, who always donned a pair of dark sunglasses and a concealed nine millimeter weapon. I jumped into the van and asked my father where we were going. He said, "To the valley." We were going on an adventure. I was very excited to finally see the valley that I had heard so much about. The valley was this magical place where my father spent most of his time. There were mangroves where manatees and crocodiles roamed, fruit trees, birds, bees and all sorts of beautiful Jamaican wildlife.

The roads to the valley were winding and bumpy. Down in the valley, I saw that my father had many lime trees. The lime trees were filled with small green lizards. I was not a fan of lizards, so I was not at all helpful with the picking process. After lime picking, I remember climbing up a steep hill. Having never seen such strange, pointy and jagged rock formations, I was filled with questions. My father explained that, once upon a time, that place was completely under the water. About halfway up the hill there stood a small wooden shack where my father kept his beekeeping and honey making supplies.

While sitting outside of the little shack, we talked about the last four years of our lives and ate delicious fried fish with hard dough bread. We washed it all down with fresh lime juice that had been sweetened with my father's honey. The view from that height was the stuff that postcards are made of.

After lunch we walked down the hill and took a dip into a shallow part of the river. I ran my toes over the many smooth rocks inside the water. My father assured me that he would keep a lookout for crocodiles. The water was very cold but refreshing.

We journeyed back up the hill to retrieve his beekeeping supplies. I watched my father tinkering with his homemade smoke making device. I observed as he added some dried herbs into one end and the smoke slowly seeping out of the other. He explained to me that he needed the smoke to calm the bees.

We made our way back down the hill again. I remained in the van parked under the shade of some trees. This time I watched my father from a distance. I remember how alien he looked in his large white beekeeping hat. I had never seen live beekeeping before, so I made sure to carefully watch his every movement. The bees swarmed around him as he walked back and forth, opening and closing the white bee boxes. The air was hot and smoky. After what seemed like forever, my father returned to the van and handed me a honeycomb. I gladly accepted this naturally golden geometrical wonder. I noticed that there were a few bees following it. I sat very still as the curious bees landed on my hands. They walked from my hands up and down my arms as if silently deciding whether I was friend or foe. Surprisingly, not scared, the bees did not sting, and I was completely calm. For the first time I felt that special connection with my father's amazing bees. In that moment, while staring into the eyes of a little bee, I understood everything.

Our Beekeeper Laura

William Hawkins

Our friend Laura had a dream. A honey bee the size of a child knocked on her front door and asked her if she had accepted honeycombs into her life.

"It's a sign," Laura said.

We told her it wasn't a sign, it was only a dream. We told her not to do exactly what she did: quit her job, surrender her benefits, leap without anything to catch her but a dream of oversized insects at the front door.

"It wasn't just *a* dream," Laura retorted.

Our friend Laura had many dreams. In one, she directed traffic at the intersection of four 10-frame Langstroth hives. In another, it rained honey, and she watched it stick to the bay window of her kitchen nook as she sipped a chai latte. Her subconscious would not cease, she said, until she accepted what it had to tell her.

And what, we asked, is that?

"I'm a beekeeper," Laura said.

We asked her if she had ever kept bees before.

"No."

We asked her if she had ever met a beekeeper.

"No."

We asked her if she was truly prepared to sink her hard-won savings into such a cockamamie scheme.

"Yes."

So we asked Joe instead. Joe the lawyer, Joe the reasonable, Joe the husband.

"Guys, it's out of my hands."

And it was. In the years after Laura's dreams, white stacks with buzzing edges appeared in her front yard. Often the neighbors would watch Laura walk among the hives in her beekeeping suit and hood. They watched through the blinds of their windows. When Laura caught them peeping, she would wave. Sometimes, she would come to their doors, ask if they would like to try a batch of her honey. They accepted her honey. They chatted, asked her how she was doing. And then they closed the door and went back to the being behind the blinds.

Laura didn't care. We knew because when she talked to us about her beekeeping—usually when she'd cornered us at dinner parties—she had a smile on her face wider than her cheeks could hold. We knew because, as the years went on, the glass jars of honey she gave us—always with a square piece of bandanna under the lid, usually a bright solid, sometimes a faded pastel—the honey got better, sweeter, deeper in flavor. We knew because her labels started out as name tags scribbled with Magic Marker and ended as professionally printed stationery. We knew that Laura didn't care what people thought of her, she only cared that the bees accept her shadow over the lives they led.

They must have, because in the years after she began, her honey appeared in local coffee shops and boutique bakeries. We told Laura, whenever we could, we were wrong. She'd been right all along. We confessed to her—usually when we cornered her at dinner parties—our own secret wishes, the expectations we'd once held of ourselves. We told her what our own dreams had long since ceased reminding us of.

And when the glass jars of honey began to dwindle, when she was confined to her bed, we told her of the sweetness she had lent to our lives. We would never have our dreams, but we had at least had the great fortune of watching someone achieve theirs. She smiled when we told her, but by then it was such shadow of what it had once been it hurt us to look on it. So instead we looked out the window, the window Joe kept open at Laura's insistence and, through it, we heard the persevering buzz of the bees as they flew in an arthropod halo around four 10 Frame Langstroth Hives. Their honeycombed lives went on, and the sound of it lulled their keeper to sleep.

"Just an Assistant"

Helen Ackermann

So the question I asked my husband was, "How did your hobby become my hobby?" When he retired from being a German teacher—after thirty-five years in the classroom—George was ready to do some of the things he had always wanted to do. He had expressed a desire to keep bees and, in his very German way, began the process of becoming a beekeeper.

The first step in his plan was to join the Marathon County Beekeepers Association. Marathon County is located in central Wisconsin, a county of warm summers and very cold winters. Attending the first several meetings, we heard of the many challenges there were to keeping bees, from mites to diseases and, of course, to cold winters when making sure the bees had enough to eat was a priority. He began to gather a number of books together and read them, studying the many aspects of beekeeping. He also subscribed to *Bee Culture* magazine and still reads it from cover to cover. He asked how I would feel about having bees. I said, "Sure," not realizing where this agreeable comment would take me. My job was to ask a friend whether we could keep the bees out on his hobby farm as the village of Rothschild where we live has an ordinance against the keeping of bees. Terry, my friend, was delighted and soon he and George became friends too. And so it began. We ordered packages of bees, two that very first year, eight years ago. I had no idea what this would entail; I was not always listening all that carefully when George told me about the bees.

This was, after all, supposed to be George's hobby, but soon I was involved quite heavily. We have the kind of marriage where many activities are shared. We both grew up on farms, have been involved in education, like to travel (I more than George), and, in general, love the out-of-doors, appreciating all things of nature. I could not help but be interested in the bees. Off we went to pick up our packages of bees in southern Wisconsin. We brought them home and then took them out to their new home at Terry's. George had prepared everything for their arrival and I helped set up the hives. We placed the bees in the hives, put in the queen cage and came back the next day to see what had happened. The bees were happily in their hives with the queen out of her cage. We were excited. And so it began.

From that point on, it became a work in progress. It was only occasionally that I did not accompany George to the bees. I did enjoy the ride in the countryside and visiting with Terry and his wife Mary. I learned to don my hat and veil, gloves and jacket. I learned to use a hive tool, how to start the smoker, how to help lift the honey frames out of the hive when honey was ready to be extracted. I found a recipe for candy fondant and, this past fall, made twenty batches. We use it to feed the bees over the winter in addition to the honey that we leave for them. I had a role in extraction as well and also made contacts with friends whom I knew might be interested in purchasing some of our honey. As an assistant beekeeper, I guess I was not so surprised when I was elected the secretary of the Beekeeping Association even though I was not present at the meeting when the election took place. I helped set up our display area for our local fair and even kept a journal as to when we went to the bees, what we did and a report on how they were doing.

I am not a full-fledged beekeeper, but pretty darn close and that is okay. We enjoy this hobby together and I think that is the most important thing. Sharing in the wonder of bees and enjoying the honey that is produced is a wonderful thing. I am not sure I would ever keep bees on my own, but for now I do enjoy the assistant role. I guess I don't mind that his hobby has, in a way, become my hobby too. After all, I love the guy.

Our Grandfather, the Beekeeper
Laquavia Foster

Erin sketched the view outside of the window using her box of colored pencils while I preoccupied myself with the corner of the room that seemed untouched by time; a nook where a dog-eared edition of *The Joys of Beekeeping* laid beside framed photographs of our grandparents, a notebook filled with pressed flowers and other mementos that held the faint, fading scent that could belong to no one else but my late grandmother. My fingers traced the frames, then the round, grinning faces of my grandparents. So preoccupied I was, that I was startled to find my grandfather in the room, drying his hands with a checkered dish rag. "Come on the two of 'ya, before it goes cold."

I suppose it was a comical sight, the three of us—Erin who towered over the two of us like a sunflower despite being younger, grandfather, and me, all cooped together in his kitchen. Like the Matryoshka dolls our mother let gather dust on the polished desk of her office, we sat descending in height, our elbows touching and bumping into the plates and rusted silverware before us. A loaf of rye bread still warm to the touch, sliced into blocks that we could adorn with fried sausage links, eggs, marmalade and grandfather's absolute favorite, honey. We watched as he took a knife to his slices, at how the little scar near his mouth disappeared behind the folds of a grin before he'd take a bite. At that, we too, would begin to eat, basking in the warmth the space provided. "It may be simple," He said, crumbs sticking to both sides of his mouth while he chewed. "But nothing beats honey on toast."

After we'd done the dishes, dried and put them up, grandfather would sit idle in his favorite chair—filling the backyard with the rich, robust sweetness of his pipe while the sun cast shadows across his still frame. He sat, listening to Erin and I as we ran about, hands and legs flung in the summer air; tumbling around until our lungs conceded defeat, and we found ourselves flat on our backs. We were children amok with fallen and broken tree branches, daughters of a professor and a salesman who didn't tolerate such behavior. Grandfather, however, understood us and welcomed our imaginative adventures. His rasped laughter like encouragement, carried out on the river of smoke that flowed steady from where he sat. "They're kids 'fer god's sake. You keep 'em cooped up all the time in that city, busy with lessons and whatever else. I say, let 'em explore." He would tell our parents over the phone when they'd ring to see how he was getting along. He'd disappear inside the house when the afternoon hour strolled in, only to return in dusted overalls and boots that'd seen better days, his pipe tucked away into a pocket on the breast of his checkered shirt. A green watering can clanked about at his side, sloshing water out onto the ground. "Time 'ta see after the flowers, kiddos."

We followed along behind him in his garden, listening as he identified the various plants he'd grown. Sunflowers, we knew, but lavender he identified for us, along with lion's tails, oregano, cosmos, cornflowers that swayed about in all their blue magnificence. That's where we found a few bees, buzzing and flitting around the florets. "Didya know both your grandma and John F. Kennedy fancied this flower? The bees got good taste. C'mon, let's leave 'em to it."

There was calmness, a cloak of serenity that draped itself around us as we continued on. Overhead, the leaves rustled in the streams of light and fell at our feet, and in that moment, bathed in the sun, our grandfather was illuminated; his hair became tufts of silver, and the scars crisscrossing his cheek, pink like the inside of a salmon turned white. "Do you miss it?" Erin asked, scratching at the heel of her sandaled foot where an old bandage had begun peeling and curling at the edges. I nudged at her side with my elbow, a disapproving scowl marring my face. It went without words, an unspoken agreement that we didn't ask grandfather specific questions that'd trigger certain memories. Grandfather halted in his steps, and his fingers touched at his hidden pipe as if he'd have a smoke.

"Your great-grandfather was a beekeeper, and I had my time too. I look back at that time with nothing but fondness, but the most I can do now is offer these bees a place 'fer them to eat. Doctor's orders tell me I shouldn't be doing that either, but after all they've done, it'd be awfully rude of me not to. We owe much to the bees, from pollination to being used for medicine. Bet they never taught 'ya that in World War I, they used honey on the wounds of soldiers like me, or that the use of honey has gone on further than that. So I took myself down to the library, borrowed some books and did the best I could with my two hands. It's not much, compared to what I used to do, I do admit, but I'd thought I'd help out these little guys anyway I could. I figured, it wouldn't hurt for this old chimney to do something other than blow smoke."

Grandfather scratched at his cheek, a smile tugging at his lips. "Seems I didn't really answer 'yer question, did I?"

"S'okay. Can I water the plants?" Erin asked, and he obliged, passing the can to her. Together, they stood—her bent form tipping the can over as grandfather observed, watching to make sure she wouldn't drown the plants. His words repeated themselves as I made my way over to the two of them, and stood next to grandfather. His hand came to rest on my shoulder, and he regarded me with a smile before turning back to supervise Erin. "Don't play favorites with the flowers, they all need water."

After that afternoon, he said no more about the subject, never said much when it came to the subject of his life as a beekeeper, and Erin and I never brought it up. However, if there was one thing I knew, whether he spoke about it or not, was that he loved it—more than words, or his actions, could ever express.

An Unexpected Career
Jane Resides

Ten years after moving to New England, we bought what had always been my dream, a very old house. The previous owners had used it as a summerhouse and evidently hadn't occupied it for quite a while. It was a charming cape colonial with central chimney and three fireplaces—pretty much what I'd wanted, but there was a problem. Before we could move in, the honeybees who'd moved in before us had to be evicted.

The bee colony had ensconced itself in an area under the roof at the back of the house. Fortunately, we didn't have to hire a professional to remove them because Jeff, a co-worker of my husband Frank, was a beekeeper. According to Jeff, this would be a "breeze." The plan was to remove a section of the roof, capture the bees in a hive box and collect the honey. Frank would become a beekeeper and they'd have lots of honey to share.

When the day came for them to achieve this clever feat, Jeff supplied Frank with a white beekeeper's suit and a bee hat with netting. The hive box was set up on a flat section of the roof. The colony was cut out from their self-made hive and placed in the hive box—all worked as planned. Frank and Jeff congratulated themselves on a job well done then left the area, allowing the queen to get her entourage settled in their new home.

Well, as we all know, things don't always happen the way we plan. When Frank and Jeff returned the next day, the hive box was empty. They figured the queen may have been killed during the process, or she just plain took her colony elsewhere. But that was the beginning of Frank's 35 years of beekeeping.

Though he was fully employed most of the time and usually kept only three to four hives, I don't think he has ever considered it a hobby. Every fall, he wraps the hives for winter, makes repairs to hive parts, and feeds the bees as necessary. In the spring, he treats them for Varroa mites, orders new queens to replace those that didn't survive, and then gets ready to go at it all over again.

Frank's New England beekeeping career lasted only four years. When he was transferred to Pennsylvania, the bees came along in the back of his truck. Down through the years, the method of harvesting the honey has changed very little. For extracting, supers filled with frames are always hauled to our garage where all doors and windows are kept closed to shut out ranging bees that follow the scent of honey. Since our honey is taken in August, fans are a much-needed item for the garage is stiflingly hot.

Frank's first extractor had no lid, so a sticky mist coated the hair and clothes of anyone turning the cylinder by its hand crank. Years later, an extractor with a lid, and which was infinitely easier to crank, was purchased. Now, instead of continually dipping the cold extracting knives into hot water before skimming off the wax caps, he has a heated electric knife courtesy of our daughter Leigh.

And she is not the only family member who has been involved. Our grandson, Matthew, at barely five, learned the most wonderful benefit about beekeeping when he stuck his finger under the dripping extractor and gave that year's harvest his enthusiastic approval. His actual participation came at age six when he screwed lids on jars after Frank filled them. As he got older, he helped crank the nine-frame extractor and then learned how to decap frames. Now fourteen, Matthew's name is also on the label as he does all the heavy lifting of the supers and a good bit of extracting.

With a "Honey for Sale" sign on his pickup truck and one at the end of our driveway, Frank is always open for business. Though he starts most years with three hives, sometimes he loses one to swarming or some unexplained problem where the hive is full one day and is empty the next. But he can hardly complain. He gets over 200 lbs of honey in a good year and he and Matthew once won First Prize for their honey at the Goshen Country Fair.

All things considered, the colony of bees that made us wait before moving into the old house was a blessing. I like to think that the queen did survive and that she and her colony lived happily ever after—or something like that. After all, she did us a big favor.

The Bee Boxes

Michelle Purvis

I remember being afraid of bees back then. In my defense, I was still too young to distinguish one insect from another. To me, honey bees, hornets, and wasps were all the same enemy. They had stingers. They came after my peanut butter and jelly sandwiches at the park. And they were the reason Jamie died. It didn't matter that Jamie was just a boy in a book. I still cried.

Mrs. Stockholm gathered us up around the foot of her rocking chair when she first introduced us to the story of Jamie and his friend. I think she read us *A Taste of Blackberries* because she wanted us to begin to understand that sometimes things don't work out the way we hope. Some things change us. Some things leave us sad. And some things are nightmare-scary, like the thought of dying from a bee sting.

My father seemed to love bees, which was confusing in many ways—most especially because I'd watched him being rushed off to the hospital after a run-in with an angry band of wasps. It was a humid July day when they'd surprised him. He was taking the tarp off of his 6600 John Deere combine. He was all the way to the top of the ladder before he realized what was happening. He broke his foot trying to get down. When his face and left arm started puffing up from the poison, I was afraid he'd turn out like Jamie.

It was a few years later when he put the bee boxes up on the "north forty". Bee boxes. That's what I called them. Even though I was now old enough to understand that honey bees were not the same as yellow jackets, my position on getting stung hadn't changed much. When he asked me if I wanted to go and visit the hives with him, I always managed to find an acceptable excuse for staying home. But no matter how many times I declined, he kept asking.

"I'm heading out to the bee boxes. It's a beautiful day. Are you sure you wouldn't like to come along?"

His eyes looked so hopeful. "Would I have to get out of the truck?"

"No, not unless you want to."

"Could I keep the windows rolled all the way up?"

"If you really need to."

It was settled then. At first, I kept my promise about rolling all of the windows up tight. I could tell he was annoyed when he climbed back into the truck-cab-turned-sweatbox, but he never said anything.

The north forty is a part of our 250-acre family farm, passed down through three generations of Johnsons in the years since 1901. My parents tended the fields during the weekends, after working full-time jobs in the city. I learned a lot about honey bees during the trips back and forth to the "home place," especially when my mom and brother stayed behind in the city. I often peppered my father with questions during the 45-mile journey.

"Why do you like bees so much, anyway?"

"I like 'em because they're interesting and they do an important job in this world. Plus, the honey is pretty delicious, don't you think?"

I couldn't argue with the honey part. "But aren't you afraid one will sting you?"

"Oh, sure, a little sometimes, but that's why I have my suit and the smoker. And respect for the bees. Honey bees mostly mind their own business unless you bother them."

I was fascinated by the smoker. I imagined the smoke putting all of the bees into a sort of hypnotic trance. I wondered if they looked like they were playing a game of freeze tag. Though I didn't quite grasp how the "magical" smoke worked, I liked the idea that the bees were somehow distracted when my dad was working with the hives.

"How do you get the bees to stay in their boxes?"

"Well, they don't stay in them all of the time because they need to go out and get pollen and nectar to make honey. But I get them to go into the hives by first letting the queen out of her little cage and then the worker bees after that."

"Why does the queen have her own cage?"

"The queen is very special for many reasons. She's the only one who lays eggs that become worker bees. And all of her children follow and guard her, so she is also kind of like a ruler too." I imagined myself as a special queen, ruling from a golden throne with everyone trying to protect me.

We often journeyed along in silence until more questions occurred to me.

"Why do you give them food and medicine? I thought they got their food from the flowers."

"Well, sometimes it's too early for them to find food on their own, so we help them a little. And sometimes bees get sick, but the medicine helps them fight disease, kind of like when you go to the doctor to get shots." A bee getting a shot. That seemed a little funny.

***

I liked harvest time the best. Dad would skim the hot wax-knife over the top of the frames before placing them into the extractor that was set up in our basement. I tried to be interested in his explanations about centrifugal force, but mostly I wanted to take my turn with the crank. And when it was finally time for tasting, my brother and I got to have a bit of the comb, all beautiful and drenched in honey. We chewed the wax until it disintegrated into tiny slivers on our tongues.

I was in sixth grade when my dad and I enjoyed our last honey bee moment. Two other things happened that year. My dad earned a promotion, and we moved to California, over 1200 miles from our family farm. Before we packed up and headed south, my dad arranged to be a guest speaker in my science class. To prepare for his presentation, he built a rectangular glass case with just enough clearance to slide in one bee-filled frame. He drilled small holes to make sure the bees could breathe and attached a brass latch to keep the case closed once the frame was inserted. Despite his discomfort speaking in front of others, he answered my classmates' eager questions for almost an hour.

A few months ago—after watching *Vanishing Bees* on Netflix—I prompt-ly went to my local nursery and purchased 86 dollars' worth of perenni-als, including two fuchsia-colored bee balm plants. I knew it would be a while before the bees sensed the newly-planted flowers, but I sat down in the dirt and waited just the same.

Hive Mentality

Catherine Blakely

Last May, I decided I needed a bee hive. I wanted bees. God knows why, although my old age is due to begin next Tuesday, I'm bored at work, and am in the midst of a mid-life crisis. I liked the idea of joining a tradition, of learning an old skill set, and bees interest me. I'm sure there's a metaphor for something in here. Probably my life.

This isn't about honey. I go through less than two small jars of honey every year. I don't care about selling honey. I have no particular interest in honey other than an occasional spoonful in herbal tea, one recipe for black sticky gingerbread that calls for it, and toast with honey as a late night snack when I can't sleep

But, as it turns out, I was too late. Too late in the season to give a hive a decent chance at survival. "That's good," said someone, "You can spend this year learning about beekeeping." I knew she was right, but that's not how I wanted to do it and the year stretching out before me made my neck tighten up.

Spoiler alert; still no bees. It's November now. I read bee books and magazines and go to bee association meetings where speakers say things like, "Americans like a yellow bee," and discuss class A, B, and C noxious weed elimination versus bee forage needs. I learn that my favorite plant, with its dinner plate size leaves, is called knot weed. Last meeting, an audience member got whipped up about non-native plant species and spoke out of turn to the woman lecturer.

"No! No! You're getting it all wrong!" he practically shouted. It was very dramatic.

Then he got a hold on himself and tried to explain. He became almost teary lamenting thistle weed spread in eastern Washington, or something. Personally I don't think he would have spoken like that to a male speaker, and I wanted to say so, but I'm new so I didn't. Plus, who asked me, anyway? One man piped up, "Well, bees aren't native either," and the meeting moved on.

I thought bee association meetings would consist of five old guys in shirt sleeves. What I got was a meeting of over a hundred people in a junior high cafeteria beneath a Beyoncé *Got Milk?* poster. The five old guys were there, though, seated together and chatting throughout the entire presentation. Halfway through, there was a twenty minute break for socializing, with drinks and snacks at the back table.

I felt the old timey, insidery, rural grange hall meeting vibe that I'd hoped for and it warmed the cockles of my heart to see all these people at cafeteria tables on a Monday night earnestly discussing regional honey prices and books with titles like *Swarm Essentials.* An older man stood up (in plaid shirt sleeves), bursting with pride, and thanked all those who'd helped him with a problem the previous weekend.

"And I successfully handled my first swarm," he announced. Everybody clapped.

Aristotle, Tolstoy, and my Canadian grain farmer uncle, Bob, were all beekeepers. Uncle Bob sends his handwritten lecture notes from the 1950's beekeeping portion of agricultural college, smelling like basement and bound together with safety pins. They are written with the kind of pens we wrote with before ballpoints, back when people had good handwriting and used blotters. I loved reading them up until the section on wintering over, or rather, not wintering over bees in sub-zero Saskatchewan. They advised a teaspoon of cyanide powder on a square of paper placed in the hive entrance. Then block the entrance. Worse than this post-harvest, ungrateful slaughter, they simply buried the whole toxic mess in the yard.

I look up translations of the word beekeeper in other languages, just for fun. In Spanish it's *encargado de la abeja,* in German, *imker,* a French beekeeper is an *apiculteur,* a Swedish one, *biodlare,* Italian, *apicoltore,* Dutch, *bijenhouder.* Did not find one in Icelandic. Are there no bees there? Is it too cold?

The Bible, Torah, and Koran all mention bees. A 256 B.C. Egyptian papyrus describes a large scale beekeeper. Zeus turned nymph Melissa into a honey bee. Monks kept hives in the Dark Ages.

Moving right along, we jump to North America where settlers introduce honey bees and the Native Americans call them "white man's fly". And the beat goes on. Our way forward has been fueled by honey and lit with beeswax candles. Surely we can assist them through a cold winter in Saskatchewan?

Uncle Bob asks, "Can you trust them?"

It takes me a minute but I gather that what he is asking is if I am afraid of bees. No, a bee sting is not the kind of pain that worries me. I remind him of the time he took me fishing—he carrying the rods, me with the tackle box, bush-whacking a trail down to the river. I became entangled in brambles and, when I looked down to untangle myself, I found bees all over my hands and immediately started screaming. I threw a fifty yard pass with the tackle box and ran shrieking back to the car.

He snorts dismissively and says, "Oh, that was just them wild bees."

We think we invent everything. The drama of bee life is no less enthralling than a Hollywood plot arc or Greek tragedy. I think of *Game of Thrones* mixed with the British monarchy. It's all there; ruin, sex, hunger, scrounging for food, family, home invasions, fratricide, survival of snowy winters, God, ecology, productivity, and working the family business.

A queen mates in one flight and is escorted to the entrance of the hive for this flight by escort bees. Male drone bees only exist for sexual participation in this one flight. A hive is populated by thousands of sisters. Guard bees will die to protect the hive, attendant bees groom and feed only the queen. Should the queen weaken, nurse bees will immediately begin to raise multiple new potential queens in queen cells. Incapable queens are overthrown in what is called a supersedure and beekeepers talk of a distinctive roaring sound in a queenless hive. The first new queen hatched makes a piping sound that the others respond to from within their own queen cells. She then moves along and kills them all in their beds before they can hatch and usurp her.

Are you a drone? End of the season? Sorry, we can't feed you this winter. You'll be taken to the front door and given the boot, wings chewed off, unceremoniously pushed over the edge. Wait, there's more! Chains of bees holding "hands" to form plumb lines and scaffolding as hives are constructed in tree hollows, bee waggle dances and democratic process and canvassing when foraging for food and new homes. I'm enthralled. There is no getting away from it; I must have bees.

I'm thinking of these magnificent natural things as I make my bi-annual honey purchase at a small local co-op where an entire tapped, green barrel lies on its side, labeled "Raw Blackberry". It's a weekend afternoon and I'm wearing my pajama pants under a big coat. I live in a place where you can wear pajamas around town and no one bats an eye, because it's the west coast, and that's how we roll.

I squat down to fill a container I bring from home. I purposely only open the tap a bit so I can watch as it slowly unravels itself into the new space, and this is what I was doing, zoning out, watching ribbony honey patterns like some big old stoner in pajama pants, when I realized that there was a small child squatted closely beside me, observing.

He said, "What are you doing?"

"Getting honey," I answered.

There was a good five-Mississippi pause.

"I like honey," he said suggestively, as if it were a code phrase, as if we were two spies in a John LeCarre novel about to exchange top secret information. And, as it turned out, we were.

"Put your hand under there," I instructed him, and then ran honey into his cupped hand. Isn't honey always stolen?

I don't know what your personal constitution is, but an improvisational event like this can complete a day for me. It might even make my week. Woman in pajama pants filling honey jar. Grinning child maneuvering entire hand covered in honey into his mouth. Two people squatted in a grocery aisle, breaking the health department rules, participating in a long tradition of people, bees, and honey.

WAXING

Bastion of the Bees

Rebekah Hoeft

If they knew
that honey, gold,
would be, to their dismay,

pilfered from
their timbered keep,
I wonder if they'd stay?

Would they wing
and find a place
where humans could not go?

Citadel
in hidden trees
where treasure sweet could flow.

Or…

…do they know
and count the theft
as tribute well-deserved?

Just reward
for room and board,
for care which they are served

gently by
a knight in white
who calmly tends their hive;

takes his due
and leaves the rest:
ensures that they survive.

The Melt of Honey

Bill Pendergraft

We are in space,
turning like bees around color
gathering and returning to build
to and from our tasks
as straight as sunlight.

We do not bend or waiver.
We do not know of this or that.
We go straight to it
following some memory,
something we heard our mothers whisper
as we fell upon our sleep.

Occasionally,
so seldom that it makes me wonder
if I have thoughts more than bees.
I ask myself why I fly
and what my buzzing serves.

I have harvested honey.
I have smoked the bee,
and in his lethargy,
stolen all he meant to keep.

It is a dream
not of keepers or even bees
but of the milky sweetness of the work.
The melt of honey on the tongue.

Through the Window
Zosia Botto

Yellow rays of sunshine bled through the drapes, and the window had been raised high all through the night. I was, in fact, glad to have the crisp morning air wandering in, raising goosebumps along my arms. There was no need to fear anyone stealing anything. My place was far out from any other life form except the deer which took residence on my property during hunting season. It was well known around this area, if a deer was on my land, you leave him be or you'd find yourself in a lot of trouble. As I peered out my window, I saw a pair of bucks locking horns and, in the distance, a dainty doe, complete with big alien eyes. There was no cause for interference, so with a hot mug of coffee in my hand, I intently watched. I stood at that window for quite some time, until the smaller of the two retreated and the larger male triumphantly trotted over to the innocent doe. She nuzzled his neck and they walked off, but before they were out of sight, the doe turned her round eyes to the smaller male who made eye contact. If not for a second, they stared at each other in a small eternity before she was ushered away by the larger male while the other turned around and, too, departed.

Not sure whether to feel sad and pitiful or satisfied and lighthearted for the deer, after extensive examination decided the feelings differed from each deer. While one might feel proud, the other may feel sadness and longing. However, what did I know? I was only human. I didn't understand the mechanics of deer logic or communication. Who was I to assume anything? I was a little bee amidst a world of flowers, here to serve my purpose until my fleeting life just disappeared. After discovering I had forgotten about my coffee while lost in thought, when I brought the mug to my lips, it was cold so I poured it down the drain and continued on with my existence. Getting dressed and brushing my teeth was my favorite moment of the day. That unique experience of being perfectly clean for just one moment. It's not quite rebirth, it's something deeper, a flicker of hope. Today could be the day you discover a million dollars, make a scientific breakthrough, read a book that changes your life, or just simply the day you got up and saw a flower blooming.

There are so many flowers, all resembling each other, but each had its own flaws. One has five not six petals, another a slightly different shade of orange, and another has a bee relaxing on the upside of its petal. This bee also resembles too many others, but its pattern isn't quite the same; his hair is a different length and color than the others. All these differences and flaws help distinguish one from another. That flower from the other one.

So when that bee flew from its resting position on that orange lily, I didn't move or make a sound. The world was perfectly still for just that one moment. I knew of my allergy of bees, but she wouldn't hurt me if I didn't provoke her. She waddled up and down the palm of my hand. I counted her stripes and noted her crooked antenna. The gears turned within her mind, and in her eyes you could see her wish to fly into the air and go to wherever she wanted to go. As predicted her translucent wings began cranking and she became lighter than a feather. Without blinking I observed how she levitated from firmly standing on my hand to lazily hovering. Once again she made up her mind and with purpose fluttered into a flowerbed. After she'd disappeared I crouched down to smell the sweet scent of the lily before moving along.

The day was hot, and the day was long until the sun sank beneath the waves of grass. I pushed a loose curl out of my face and I watched the sun slink off to bed. Deciding I should do the same I gathered my work tools and walked towards the two story cottage. If not for the moon that brightly paraded across the sky I would find myself face first in the grass. After a hot meal closely followed by a scalding bath I retired to my bedroom and unlatched the window. With the light of my bedside lamp I read. At first I heard a faint buzzing sound, curious I placed my book on the nightstand and found a bee, with a crooked antenna and exactly four black stripes. Her mane could compete with a lions, but the most enchanting were her eyes. Almond shaped and glossy, similar to the doe's I'd seen that morning moved me. They spoke without words, without a sound. I struggled to understand and then I realized her stance was off. Instead of proud and graceful she was slouched. Finally spotting the injury to her leg I gingerly picked her up ignoring the fact her stinger was armed and poised and place her in a small container. In my ratty old t-shirt I rushed downstairs and squeezed a lemon in a jar, dropped a spoon of sugar, and a spoonful of water. As I mixed the concoction I walked back up the stairs and used my finger as a dropper and let a few drops fall to the bottom. She lifted up her head to look at me and I sighed. I wormed my way into the covers and fell asleep.

The next morning I stretched my arms and looked over to the container on the table. It was empty. The concoction was gone and so was the bee. I smiled to myself and I got into the shower. Maybe I'd see her on that orange flower, and I'd know exactly where to look and exactly who I was looking for.

Among the Clover
Carrie Montgomery

Barefoot and pink-polished,
I slide among the white globes,
Sweet scents ascending as
Flower heads snap from stems,
Beaming baubles
Bobbing between my toes.

Striding back and forth—
Watering can, faucet, garden—
Scanning reflexively,
I halt
Realizing what I don't see:
The winged sweet-potato bodies
Of the honeybees.

Their absence is silence.
Clover-walking rendered innocuous—
No stingers to sidestep,
No coworkers in the garden to spy on.

I think of the humans' prophecies,
Predicting the demise of honeybees.
My mind wanders to my past
Neighbor's apiary. Could I fearlessly
Mother a hive? (Yet wonder,
Is any mother ever fearless?)

Then I see her. Alone
Like me—head in clover, busy.
Maybe she notices
Her solitude,
Works harder for it, or

Maybe she is
Driven by fealty,
An ancient urge to forage, feed
Pollinate her pleasure
Upon which
So much depends.

(Upon which
We all depend.)

Truth

Vicki Wilke

She knows my fear and still
hovers in taunting hum,
glistened wings on her brazen way
to salvia, poppies, my pink rose.

She reminds me how I love
the idea of bees, pleasure of honey
and muted yellow, royal jelly,
allure of pollination.

And yet, I tense and wave,
not brave enough to let her land,
take the sting, that truth
of every beautiful thing.

I Don't Think That Much

Nathan Cox

A golden honeycomb buzzing by each ear,
    Hexagons and polygons,

I hope to interpret their voices.

Each one drones vibrations to each other in an endless series of
emotional communications;
They hush till anger and a swarm of silence,
    Broken again by the constant tickling hairs,
        Resonating,
            Ever outwards,
        Form the hive,
    Of twitching tendrils,
Pulls out each glob.

To go inside is glazed walls of caramel honey,
    Stripes to each end,
        Without knowing,
    Clicking on the eyes,
They see each casing.
The walls are burnished in fragments forming a pattern,
    How do they know?
It is their lifeblood,
    This hive,
It is a haven…
            Always.

A constant cycle of life and death is playing in the larva sanctioned,
    Translucent bean,
Rooms.
A gloss of sweet over the shape to give form,
    They are distinguished,

Each from the others,
    By the tone of the cell,
And one can name them according.

Their energy is constant,
    Behind it there is sound,
        Voices from the,
    So true,
It brings the universe together.

One with the bees,
    They speak through their wings.
One with the bees,
    We live through their nectar.
One with the bees,
    The wax is their self.

Behind the screen houses the eternal grinding that will not cease,
    The bees' channel,
        And bring forth,
    Sinew of life,
The pattern they have.

They are intelligent,
    Harmonized.

The bee doesn't hunt,
    It does not harm.

We harm the bee,
    It has so much to tell.

The bee tells us we must give to take,
    I am the flower and honey.

The bee tells us that we must communicate,
    Sing the tone they carry.

The bee shows us that we must cooperate,
    Hum a familiar tone,
Sustain our ties.
The bee shows us an intangible strength that digs far beyond what it
means to be.

So delicate and so feared,
    Why must they be neglected?
They treat each flower like we would our own child,
    Does this not say we have to learn?
Together we must buzz and hum,
    We must stay close and together,
        When we are apart,
            We die out and suffer,
        The hive mind is not evil,
    It is one to understand,
Fiercely understand.

To not understand is to die.

They do as we,
    Till they fight,
And seen then they are pests.

Pollen…
    Pollen…
        Pollen…
            It is dust in my mouth.

Bee-Whiz
Jordan Link

The fairground is an explosion of colors. Women in rainbow dresses twirl about the stage, and a torrent of multi-colored confetti, like a tropical rainstorm, showers down on carnival-goers. I've never before seen rain, though, that brought gleaming smiles to so many faces.

I turn the fruit over and over in my hands. It's quite like an abstract painting; a canvas of splattered reds, oranges, and yellows.

"Where did this come from?" In response to my inquiry, the merchant crosses his arms, and taps his foot to music only he can hear.

"Bee-Whiz. Where did you figure it came from? Its headquarters are right over there."

I furrow my brow. The peach, spurred by a nervous habit, leaps from my palm, then returns to it. Bee-Whiz is the largest self-pollinating corporation in the world, though there are others concealed in the shadows of nearby mountains and in forests of rolling wheat.

"Are you going to pay for that?"

"I'll pass." I return the peach to its stand and continue to wander, faceless and anonymous as a herring in a school, through crowds of revellers.

An hour elapses and I realize I've spent the bulk of it eyeing the side door of Bee-Whiz. It opens in thirty minute increments to permit the entry of fresh supplies and laborers. Before my mind can persuade my heart otherwise, I near one of the site's workers. Her hat dangles from the back pocket of her drooping, denim jeans. I am no thief and am not, by any stretch of the imagination, worthy of the descriptor *clandestine*, but curiosity compels my hand forward and takes the necessary measure to satiate its hunger.

I loiter outside for a few minutes, slip the employee's cap over my already-ruffled hair, and plunge into the factory. The smell alone, an unholy potpourri of sweat and humidity, assaults me. I stagger into a pyramid of taped boxes, which, to my dismay, line each wall. Workers with glistening brows and stopwatches man assembly lines at the center of the facility and fluorescent lighting casts the spectacle in superficiality. The moment before I convince myself that my suspicions concerning Bee-Whiz were unfounded, I spot a behemoth sign. I approach the heavyset woman who stands beneath it. She unwraps and dissects me with angular, turquoise eyes and marks something in her clipboard.

"I would like to access the grounds, please."

"You're going to need your beekeeping suit. Make sure you have your gloves and that you've inspected your veil for holes." She ushers me toward a pile of white uniforms stacked in a single plastic bin against the wall. I shuffle over and snatch the first one that catches my eye. Luckily, I share the true owner's build.

I follow her into a conservatory of sorts and the world cracks open before my eyes. Lies, like honey in consistency, flow freely from the carcass of the truth I thought I knew. The stone walls of the garden obscure this haven, this Olympus, from the scrutinizing eye of the mortal media. Its trellises lean so far forward that I fear they might expunge me.

Something zips past my ear; a dream, *no*, a nightmare.

We turn the corner, and I scramble to gather my jaw from the floor. Apiaries, as grandiose of Grecian statues, loom over me. A soft song emanates from them, the ballad of honey bees hard at work.

"The last ones," I whisper.

"That's right."

We stare in silence at a scene so few have witnessed. Then, some higher being places a thought in my mind, a thought that no amount of repression can uproot.

"Hey, do those gates open?"

"There's a button over there, but it's only for emergency situations that could result in the loss of our bees."

"Your bees?"

The worker shoots me a quizzical look, and shrugs, a tad defensive.

"I will never fully articulate my gratitude."

I venture over, inhale, and slam my palm into the button. An alarm wails and quells the monopoly's assiduous symphony. Hundreds of security officers, shouting slanderous words, flood towards me and the doors. Though they lug me, a new prisoner, into the darkness, inch by fighting inch, balls of golden luminance in the form of pollinating insects find freedom. In an instant, a group that resembles an army battalion rushes past with an armory of nets and nucleus boxes.

"You misstepped today, son."

A fit of laughter seizes me. From now on, children will run along summer grasses barefoot, and scale trees to retrieve their ripening gifts. Fields of sunflowers silhouetted by fog-laden blue mountains will, once again, become the faces of happiness, of freedom.

"Do you really believe that?"

The Idea of Bees

Vicki Wilke

I love the idea of bees, castles steeped
in honey, drones, workers, a bulbous
queen—courtly brood, knowing her well,
her smell.

How perfect, six-sided cradles, larvae
swaddled in warm wax, devoted nurse
nannies, all the plundered pollen
they can eat.

I love the idea of bees busy from birth,
adoring their queen in a cooled hive, filling
fleshy pollen baskets, bloom dust stuck
to static legs, homing in on a hum.

How sultry, pampered queen I'd like to be,
plumped and coddled with royal jelly,
dining until death, carried away,
like a sweet bouquet.

I love that they dance sole and brave,
wiggle in a whirring blur, point other
amber heads to nectar fields, errands
on diaphanous wings to feed the earth.

How zealous, worker girls, defend,
clean and repair, like the soulful mothers
they will never be. No busybody dawdling,
only beelines back home.

I love that they swarm, cluster their queen in
a massive loyal pile, stripes and hums
bumbled together, tuned for a brazen scout's
waggle—perfect new home.
I just love the idea of bees.

Bee Attitudes

Mary Langer Thompson

And seeing the black barrel buzz outside the window,
the teacher opens her mouth and teaches her students, saying:

Blessed are the poor workers, for theirs is the kingdom
of honey and sweetness.

Blessed are the colonizers, for they shall gather together
in true democracy.

Blessed are the meek, for they shall not provoke
or resort to stinging.

Blessed are they who hunger and thirst after nectar from the fields,
for they shall be filled.

Blessed are the mercy keepers, for they shall assist
with temporary homes and shelter.

Blessed are the pure of heart, for they shall not use
or be destroyed by nasty poisons.

Blessed are the peacemakers, for they will help workers
reach the flower patch.

Blessed are those who are persecuted,
for they shall transform nectar into honey.

Rejoice, observe and learn the waggle dance,
for by it you will gain direction, lead the others, and be glad.

The Stadium of Bees

Anna Citrino

We had climbed Delphi's stone-strewn road
to step into the ancient stadium, nestled into the hill's
height surrounded by bay laurel and pine, thick

with yellow pollen. We stared down the field's
enormous length, surveying the multitude of seats,
imagining the crowd's clamoring cries bursting in bolts

under a burning, blue sky and Pythian competition heat.
But what we heard instead as we stood gripped with stillness,
listening to the sun-sharpened air for what once might have been,

was a low hum rising. Animated and alive, the stadium air
vibrated with the ten million bee wings whirring,
the bees' engine thrums resonating against the stony seats.

Air thick with pollen spice, flecks drifting down
to dust the earth below, we breathed in—burying our bodies
in the scent and sound, holding it as if it were honey.

We had come from a desert world filled with piercing
white light, and sea bed that had lifted itself from the ocean
millennia ago. For minutes, we stood there melted

into the murmuring, our ears enormous amphorae,
receptacles for a universe of flowers where bees fly in,
legs swollen and weighted with pollen, the hemispheres

of our skulls flooded over with the sound of aum.
There are deserts in this world, places where no music
sings—the sea a distant memory.

But here, at Delphi, bees have broken open
the hive's heart, and humming, have carried the world
back into a cradle of wonder.

NEW WAX

A Honey of a Conversation

Billie Holladay Skelley

When I was much younger, my family had a friend, Mr. Hall, who was an avid beekeeper. It was fascinating to me to watch him, usually from a distance, as he worked, inspecting hives and harvesting honey. I was even more captivated when he talked about his passion for beekeeping. A tall, muscular man with brown, leathery skin from years of working outdoors, Mr. Hall could be quite serious when he outlined the importance of pollination, and there was a hint of pride in his voice when he detailed the benefits of honey. His self-described "bee etiquette" was somewhat amusing and I eventually concluded this etiquette was basically his protocol of safety precautions for not getting stung by his friends. What impressed me the most about Mr. Hall was that you could tell, in both his actions and his words, that he truly admired and respected his honey bees. He knew a great deal about them and he did his best to enlighten me regarding their merits. From him, I learned about the importance of honey bees for our agriculture and economy. I was made aware of the numerous nutritional and health benefits of honey. In just a few conversations, he convinced me that the little honey bee has had a huge impact on our world and our lives. Because I knew little about these insects, I could not contribute much to these conversations and, for the most part, they were one-way affairs. He talked, and I listened. One day, though, I did manage to control the conversation—at least for a little while.

"Have you ever considered," I asked Mr. Hall, "how big of an impact your honey bees have had on our literature, language and communication?"

My question was met with silence. I could tell that he had not thought about honey bees in terms of their contributions to our methods of communicating. Being more the indoor-type and an avid reader, I told him it was something I had thought about often.

"I'm constantly amazed," I said, "at how frequently I encounter references to honey bees in literature."

"What do you mean?" he asked.

"In books," I answered, "there are many examples, such as Chaucer's *Canterbury Tales* where he writes, 'for aye as busy as bees are they' and in Shakespeare's *The Tempest* where Ariel sings, 'Where the bee sucks, there suck I. In a cowslip's bell I lie.'"

Because Emily Dickinson is one of my favorites, I also told him about her and how she often referenced the industrious bee in her poems. Dickinson wrote, "Fame is a bee. It has a song—it has a sting—ah, too, it has a wing," and, "The pedigree of honey does not concern the bee; A clover, any time, to him *is* aristocracy."

"You mean," Mr. Hall said, "like in the Bible, in Exodus where they talk about 'a land flowing with milk and honey.' I always thought that must be such a great place because consider how many bees there would have to be for the land to be flowing with honey."

Nodding in agreement with Mr. Hall's statement, I went on to explain that bees have had a much greater impact on our language than just references to them in written sources. They have worked themselves into our everyday language and into many of our common expressions. They have become a part of our culture.

For example, I told Mr. Hall that since honey bees are known to be social creatures who work together cooperatively, the term "bee" has come to be used for social gatherings where people come together and perform a common task. Spinning bees were held in colonial times, and sewing and quilting bees were popular in frontier settlements. Farm families used to gather at husking bees to husk corn. Today, many schools and communities still hold spelling bees where students come together to spell words like *apiary* and *ovipositor*.

I added that many of the expressions we use today take their meanings from honey bees or their activities. These include such common descriptions as "busy as a bee" for a person who is constantly working or occupied all the time. I tried to give other examples, such as we designate a very busy place as a "beehive of activity." People also say someone "made a beeline" for the food or another object when they want to indicate the person went straight toward something. I added that the expression "bee in her bonnet" is used to indicate a female who is focused on a task or obsessed with an issue and acts with a singularity of purpose.

I was trying to make my point and also show Mr. Hall I knew something about bees, but at this juncture, I knew my feeling of superiority was going to be short-lived. Mr. Hall had grasped what I was talking about, and I could see his mind racing.

"You mean," he suddenly said, "like when Muhammad Ali said 'Float like a butterfly, sting like a bee,' right?"

"Well, I guess," I responded. "I suppose you could include things like a 'sting' operation…you know, where the police nab an unsuspecting criminal."

Rapidly, Mr. Hall started providing other examples, such as women who wear fancy "beehive hairdos," and men who refer to their sweethearts as "honey."

Not to be outdone, I mentioned that we use the term "queen bee" to refer to a girl or woman who expects others to assume what she desires and to meet those desires without question. A "queen bee" treats others in a condescending manner because she considers herself to be a leader, in charge, or in a position of power.

Mr. Hall countered with "mind your own beeswax"—the expression we use when we want someone to mind their own business and stop meddling in the affairs of others.

Back and forth we went providing "bee" words and "bee" expressions. It became a verbal game where we each tried to think of another example before the other one could. For the first time in our conversations, I felt like I was contributing.

I mentioned that the game might go on for hours if we started including the buzzing sound bees make when they fly around a hive because the word "buzz" has become quite common in our language, and it has many different meanings.

"You know," I offered, "like when people feel as if they are flying high from drugs or alcohol, they say they are 'buzzed.'"

"It can also mean flying low," Mr. Hall said, "like when pilots indicate a plane 'buzzed' the airport."

Sensing the game was starting again, I quickly tried to make additional "buzz" contributions.

"It can also signify the noise associated with hive activity. You know, as in…the school is 'buzzing' with events. It can also mean a confusing combination of talking and activity, such as there is a 'buzz' of excitement in the locker room."

I felt like I was convincing Mr. Hall that there are many meanings for "buzz," and the word is used in our conversations in many different ways. By this point, however, Mr. Hall, as usual, was one step ahead of me. He started singing!

"What's the buzz? Tell me what's happening…"

It was the lyrics from the musical *Jesus Christ Superstar*. I had no idea Mr. Hall even knew that musical, let alone that song. Obviously he did, and he made the connection before I could. At that point, I realized, we had come full circle.

For months after that conversation, if one of us thought up a new "bee" expression or "buzz" example, we would tell each other immediately. It did not matter if we met on the street, in a store, or anywhere else. I am sure other people probably thought we were crazy, blurting out our thoughts on honey bees, but I remember always trying to have an example prepared and ready, just in case I ran into him.

Sadly, Mr. Hall died several years ago. I am surprised at times how much I still remember from his conversations regarding the importance of honey bees and the benefits of honey. Most vividly, though, I recall his fondness for his valuable, little insects and how proud he was of all the contributions they make to our lives. Many years have passed, but I have not forgotten him. Every time I hear someone use a "bee" word or a "bee" expression, I think of Mr. Hall because he was the "bee's knees"—the absolute best when it came to beekeeping.

Love Blooms
Angeline Haen

When I lay on the field's colored blanket,
ask me about sacred romance
between bee and flower. What inner call to unity draws them together.
Take to the sky. Find the flowers.
Return the gifts gathered in the brightest light
to the sanctified hive's deepest dark.
Ask me what the black light knows.
Listen to what the bees say. Creator's tiny helpers carry good medicine.
We know the nectar is there, unseen, and life buzzes by,
there all along, the sweetness collected in the spirit of the flower.
I say what the bee says: love blooms.

Of Bee

J.P.A.

Stripes and stings
Queens, not kings
And…occasionally…detached wings
These are the things of bee
Swarms and mites
Robbing fights
Trying to survive cold winter nights
These, too, are the things of bee
Pollen, propolis, wax, and bread
Nursing, foraging, guarding, dead
These all are the things of bee

A Bee in Her Bonnet

Jill Crainshaw

Pastor! Something must be done.
Old Mrs. Wickenstaff has a bee in her bonnet.
She is all aflutter,
Running up and down the church aisles,
Complaining to everyone she sees
And during the passing of the peace, of all things!
Something must be done.
She demands it.
We demand it.
There, there, now, Mrs. Wickenstaff.
Tell your troubles to Jesus.
Take it to the Lord in prayer.
Pray later, Pastor!
Something must done.
All will be well, Mrs. Wickenstaff.
All manner of things shall be well.
Lay your burdens down.
Survey the wondrous cross.
Pastor, I did.
I have a bee in my bonnet.
And you look so lovely in your Easter bonnet
With all the frills upon it.
The frills and flowers are the trouble.
Get away! Shoo!
Mrs. Wickenstaff!
I once was blind but now I see.
There's a sweet, sweet spirit in this place,
Humming beneath the cross of Jesus,
Our honeycomb.
Pastor, this is no time to wax eloquent.

She is petulant. Cantankerous. Belligerent.
People are abuzz.
Do you hear the murmuring?
Something must be done.
I speak the truth in love.
Look to the cross
And see the hive of bees
In the wall behind the baptistery.
Old Mrs. Wickenstaff—
She has a bee in her bonnet.

The Taste of Happiness

Lauren Pichard

Like babysitting the baby,
I beekeep the bees
to hear the happy hum
of my little honey makers
so that their hard work may be jarred,
soon to be bought, the bustle of business
and the buzz around town:
"The honey's for sale!"
"The honey's been sold!"
I hear the happy hum
of satisfied customers.

The sweet goodness of my honey
only comes from us,
pure and raw, just the way they like it.

Spring Fever
Robbi Nester

O, April, with its soiled
shorts, its giddy sunshine!
On the hillside, blushing purple
as eczema on afflicted cheeks,
the bees again are gathering
their golden riches, excavating
the hairy slopes of orchids,
glamorous as Amazons, with
no thought of personal gain.
Chaste as grain still fattening
in the field, they dedicate
their lives to a hive of identical
others while grubs wait,
sealed into pristine cubicles
until a sister nursemaid carefully
uncorks them. Perhaps it is
the promise of the briefest
sip of nectar, more nuanced
than the most expensive spirits
and with a kick stronger than absinthe
that fuels this species-wide philanthropy.
Or are these creatures made
more virtuous than we?
That explains it: they *are*
the angels we've heard so much
about, armed with swords,
winged, industrious,
and entirely unhuman.

Honey Haiku

Angeline Haen

Thick drop of honey—

what flower do you taste of?

All of them at once.

BITTER HONEY

Smoke

Tabitha Peyton Wood

When Alyson woke up, the air in her room smelled like rotten eggs and seemed too sharp to inhale. It was late August and oppressively hot inside the house, where Alyson's father used towels around the windowsills to keep the scent from escaping. Alyson's skin was sticky with sweat. She pulled on a pair of jeans and traded her fairy-print nightgown for a plain t-shirt. She raised the window over her bed and climbed out into the field behind the house. Here the air was still hot, but it smelled of grass and honeysuckle, and a gentle breeze made the heat more bearable.

It was half a mile to the Douglass Farm, where Alyson hoped she would find Jennifer working outside. She was not disappointed.

"Morning, sweetheart!" Jennifer called out when she saw the girl at the farm's edge. "Shouldn't you be in school?"

It was Monday. Alyson had forgotten the first day of seventh grade. Now it was nearly eleven. There was no point in going so late. Plus, how would she get there? Alyson didn't want Jennifer to worry, so she said, "School starts tomorrow."

"I could have sworn it was today." Jennifer frowned. "Well, my boys have been out of school for years. What do I know?"

Jennifer was about fifty years old. She had curly grey hair, gathered in a low ponytail. She wore a straw visor that cast shadows across her dark brown eyes. Her cheeks were flushed pink with the sort of visible health Alyson hadn't found in her own mother's face in over a year. Jennifer maintained the loose-fitting wardrobe and the soft, strong shape of a woman more concerned with hard work than with high fashion. Alyson found her striking.

"I have a little more gardening to do," Jennifer said. "But I'm going to collect honey after lunch. Wanna watch?"

Alyson nodded and smiled.

"Why don't you go say hello to the bees. We can have a sandwich in twenty minutes," Jennifer said.

Alyson had met Jennifer for the first time that June. Alyson's home was not what it had been the summer before, and she couldn't stand spending all day there once the school year had ended. So she'd gone for a walk and stumbled across Jennifer Douglass, looking somewhat like an astronaut in her big white beekeeping suit, moving carefully through a dark cloud of insects on a western field of the Douglass Farm. Alyson had stood quietly under the nearby oak tree and watched until Jennifer had finished her work. Then Jennifer had stepped out of her suit and come over to say hello. She'd offered Alyson a cup of milk and two chocolate chip cookies, cementing a new friendship. Since then, Alyson had come to the farm nearly every day.

Now, Alyson crouched next to one of the hives. There were twenty-two on the farm that day, though Jennifer owned more. The others were on loan to farms across the county. Of the ones still here, a few were kept in observation hives with glass fronts, which had provided Alyson with hours of entertainment that summer. She loved to watch the bees work. Each bee seemed to know exactly where she was going. Not a single step or imperceptibly rapid flap of the wing was wasted. As such, each hive moved seamlessly, like a single organism with fifty-thousand parts. The bees danced to communicate the location of pollen, but so many other things looked from the outside as though they were simply known. Alyson didn't understand how the nurse bees knew to care for the infants, or how the builders could create such even-sided, sharp-angled hexagons without ever being taught. How did the bees know how to regulate the moisture of honey, or to raise a new queen at just the right moment? To Alyson it seemed unlikely that all this could be instinct—that such intricate knowledge of biology, architecture, chemistry and gastronomy could come folded into DNA. It seemed equally unfathomable though, that a mind capable of *learning* all that might not occasionally over think its own miraculous existence and lose sight of its role in the system; that it might not sometimes be distracted by the easy attainment of baser pleasures. The bees seemed impervious to these mental downfalls, and to Alyson they represented everything humanity should aspire to.

Eventually, Jennifer invited Alyson inside for lunch. She made two turkey sandwiches with tomatoes, basil, and arugula from the garden. Their bread was smeared with a thin layer of mayonnaise and fruity spring honey from the farm's hives. Jennifer scooped two cups of yogurt, cultured from the milk of her dairy cows, and handed Alyson a glass of fresh carrot juice. They sat down across from each other in the bright, airy kitchen, and Alyson thought how grateful she was to have found Jennifer. Her own parents' kitchen had no food in it at all, though she thought they could still afford it if they ever bothered to go to the grocery store.

"Did you have a good evening yesterday?" Jennifer asked.

Alyson nodded. "It was good weather. I walked around for a long time. I used the money you gave me to buy a slice of pizza and a book at the general store. I tried to read under a streetlight but it wasn't bright enough, so I went home. It was pretty late when I got back and I went right to sleep."

"I see," Jennifer said. "Weren't your parents worried that you were out so late?"

Alyson shook her head.

"Well I guess you'll need to get to bed early tonight. Big day tomorrow."

"What happens tomorrow?" Alyson asked, excited, though she wasn't sure what could top seeing her first honey harvest.

"You have school tomorrow," Jennifer answered.

"Oh. Right."

"Do you have a way to get there?" Jennifer asked.

Alyson shrugged. "Maybe the bus will come."

Jennifer furrowed her brow. During these sorts of conversations, she understood how fine a line existed between being helpful and pushing the child away.

"Sweetheart, school started today. I'm certain of it. They were just talking about it on the radio this morning. If the bus didn't come it's that you aren't registered," she said.

Alyson stopped eating and looked down at her hands, which sat on her lap in two tight little fists.

"Don't worry," Jennifer said. "I can help you with that this afternoon. First let's enjoy the rest of our lunch. Then we'll go outside and you can see me harvest some honey. How about that?"

Alyson looked up and smiled a little.

When they finished eating, they went out to the shed. Alyson helped Jennifer into the beekeeping suit. When Jennifer opened the hives, Alyson was to sit by the same oak tree under which they first met. It was a safe distance, but she would still be able to see. Alyson hoped that soon she might become a beekeeper herself. She wanted to be a little closer to the bees—a little closer to order.

She would also have liked for her parents to see the bees. Perhaps the insects would inspire them. When she invited them to spend an afternoon at the Douglass Farm (Jennifer had asked to meet them) they were sitting together on the couch. "We're busy, Alyson," her mother had said sternly, holding a lighter and a glass pipe. "Go and play outside." The two parents had stared at Alyson with the emptiest eyes, and she had swallowed back the fear she felt for them lately. She wouldn't broach the subject again. "Perhaps they'll come around," Jennifer said the next day. As the summer wore on, that reassurance turned increasingly hollow.

Now Alyson zipped up Jennifer's suit and checked for gaps. "Don't the bees get angry when you harvest their honey?" she asked.

"Of course," Jennifer answered. "I don't like to take too much. The bees will still have enough to eat this winter. But they don't know I'm on their side."

"Well then how do you get them to let you take it?"

"I use a little bee smoker," Jennifer said. "It doesn't hurt the bees, but it makes them less aggressive. They think there might be a fire, so they eat as much honey as they can. That way, in case they don't make it back to the hive, they'll survive until they can regroup. The eating distracts them, and being so full makes it hard to sting."

They lit the bee smoker together, so that the first hive was surrounded by twisting white clouds. Then Alyson went over to the tree and Jennifer opened the hive. Some bees still came out and fought to protect the fruits of their constant labor, but most of them turned passive. Though Alyson understood, some small part of her still ached as she saw that even such noble animals could be hypnotized this way, under the spell of a cloud of smoke.

Song of the Swarm
Elizabeth Christie

They were loud.

Sometimes, if I could keep quiet enough, I could hear them from my bedroom. Though once I heard their faint hum in the distance, I couldn't unhear it. It was stuck in my head, like the sounds of the Earth. It was supposed to be there and, if it wasn't there, then my world just seemed unbalanced. It was the sound of comfort—not overwhelming, but not so distant that it was indistinguishable from a breeze. It was clear as a soft-but-poignant bell and invigorating as a black and yellow kazoo. The bees had been my minstrels since the day I was born.

As a child, I would play a lonely game to see how far away I could go and still hear them. One time, I could have sworn I could hear them from the old oak tree on Sullivan street, though now I realize it was probably just a hopeful delusion.

And though the bees were my friends who sang to me, and played games to me, I had never seen them. The tall fence the neighbor kept around the perimeter of his property had made sure of that. I had never seen them, so I often worried if their sounds were actually cries for help— cries of loneliness because they didn't know that I was just on the other side of the fence playing their game.

My sisters would tell me that a man in a spacesuit would go and play with the bees. I didn't believe them. Why would a spaceman play with bees? There were no bees in space. Unless… unless they weren't just bees. Unless they *were* space bees! My curiosity got the better of me and one day, when I was brave enough, I spent my day outside. I played as close to the fence as I could without my mother seeing. When I heard the neighbor's back door open, I dropped my jump rope and sprung into action. I climbed to the top of the rickety old shed in our yard and, for the first time, caught glimpse of the bees in the neighbor's back yard.

There were bees. I mean, I knew there were bees, but I had never seen them. I'd only ever heard them and had good faith in my mother's answer when I asked her what it was that made that noise. There were bees. Now I could see them. Now they felt real. The weren't as brightly colored as I'd seen in cartoons. They weren't in hives like the cartoons promised, they were in boxes. They lived in boxes and the only reason I could see them now was because the man in the spacesuit had opened one of the boxes.

I watched him. It might have been an hour. It was probably four minutes. But I watched the man in the spacesuit pull out trays of bees from the boxes, like racks of cookies from the oven. Every time he did, a few would escape, but he didn't seem concerned… Not that I could see his face. I couldn't. I wasn't sure that I wanted to anyway. What if the spaceman himself was just a giant bee who lived next door? The moment I thought that, I was terrified.

I turned as quickly as I could to get off the shed before the space-bee could see me. I turned so fast that I broke the ledge of the roof. It snapped and I dropped to the ground, landing on and breaking my arm.

At the hospital, I confessed to my mom why I had been on the roof of the shed to begin with. She was someplace between mad and baffled. Maybe she was annoyed. Maybe she was concerned. Maybe she just thought I was an idiot child. All of those would have been valid.

"Why would you do that?" She'd asked, exasperated.

"I wanted to see the bees." I left out the part about the spaceman.

"You're allergic to bees!" She'd yelled.

"I am?"

"Yes!"

"Oh." I sat back and cradled my casted arm. I don't know how I had gone six and a half years without knowing that. I probably did, and just wasn't listening when I was told. "I still like them, though," I muttered. My mom looked at me, utterly exhausted by my naivety.

Mom brought Quig over after a week of me being stuck inside, driving her nuts.

"Quig," Mom explained, "is our neighbor. The one who owns the bees."

I took a good look at him—an old man, with no hair left. Definitely not a bee. Probably not an alien, but you never know. Mom had told him what had happened and asked him to come and teach me something so I didn't rot my brain away by watching too much TV.

And that's how it began. Once a week for the rest of the summer, Quig would come over and bring pictures and books, honey and wax. He taught me everything he could about his bees. He even let me name the queens. They were his pride and joy and he seemed grateful to be able to share that with someone. He lived alone, after all. His queen was long gone.

I asked him, every so often, if I could come over and watch him work. He'd have to remind me with a melancholy smile that I was allergic. I don't know… I guess in my head, I thought that maybe if I asked enough, it might not be true one day and he'd say "yes." But that never happened.

School started back up and I saw less and less of Quig. Still, I'd lay in my bed and listen to the humming across the fence. I'd leave my window open to make sure it wasn't muffled anymore than could be helped.

I did my science fair project on the different species I'd learned about. Quig even helped me pick out the pictures I used. He gave me a few sampled of his homemade products to use as well. I got an A.

Then, one day, there was an ambulance outside of Quig's house. A week later—a moving van. No one ever told me. I guess they didn't need to. One day, I laid in my bed, closed my eyes, and listened as well as I could.

Silence. The air was empty and lifeless.

The bees were gone. And I couldn't sleep.

Keeping Watch

Dan Stout

Lacie only kept one hive despite all the books and websites advising against it. They said keeping a single hive was rolling the dice. But she kept just one, and she tended to it well. She mentally ticked off her hive tasks for the day as she dressed, only pausing when she looked down the hallway and saw the door to Jenny's room. It was closed, of course. It had been for almost a year. No one had been in the room since the detectives had visited. Tall men with condescending voices who told her not to worry, that sometimes teenagers run off for a joyride and return in a few days.

The same detectives were back a few days later, this time talking about alcohol in the bloodstream and how sorry they were for her loss. It had taken everything she had to resist smacking the false pity off their faces.

Realizing she was staring at the closed door, she forced herself to stop dwelling on it. It was just a door, but to open it would be tearing open that wound. Even when the acceptance letter with the maize and blue University of Michigan logo arrived shortly after the funeral, Lacie had simply slid it under the door.

"It's my life. I'll say goodbye when I'm ready," she told herself.

Lacie donned her protective veil. She wore a long-sleeved shirt, gloves and the cuffs of her loose pants were tucked into her socks. She always wore the gear when she worked, even on hot summer days when the air practically dripped with humidity. The uniform was uncomfortable, but the stings were more painful still.

Out in the yard where the property abutted a neighbor's fallow field, the hive sat near a tree. It was close enough for the broad trunk to provide a wind block and on the southern exposure to minimize shade. Lacie lit the hand-held smoker and pumped its bellows as she approached. The sweet smell of tobacco infused the air, and, as it reached the bees, they retreated into the hive.

The buzzing went quiet quicker than normal. Lacie frowned and opened the hive's top, revealing its extraction rows.

"What's going on with you guys today?" She peered inside and looked over the huddled bodies of her wards.

When she saw the reduced number of bees, she tried to stay calm. She poked and prodded, disturbing the hive far more than she normally would. She moved the queen excluder, which kept the queen and her drones from wandering, and looked into the heart of the hive. As designed, the queen and her mates were there, but far too few workers surrounded them. Lacie gently replaced the top of the hive shed. She turned off the smoker, and realized that she was about to be sick. She broke into a run.

Scientists call it "Colony Collapse Disorder". It's a sterile term for bees abandoning the hive. When a hive is stricken with CCD, the queen's children abandon her, flying off into the dusky night with no chance for survival. The queen is left to die alone, and for months even other swarms and wax moths will avoid the sad, empty shell of the hive.

Lacie stood in the bathroom, rinsing her mouth out and doing her best to make eye contact with her reflection in the mirror. When she walked out, the door to Jenny's room loomed even larger down the hallway. She had a brief impulse to go in, to search the photos and stuffed animals for... what?

In the back of her head, Lacie's own voice answered: *For the reasons children leave their parents.*

But she didn't open the door. Instead, with her hand on the doorknob, she bit her lip and kept the tears at bay. After a long moment she straightened up and went to make some tea.

Tea had once been a social event for Lacie, but after Jenny was gone, having company became a struggle. Her friends looked at her differently. *Poor Lacie*, their eyes said. *First a widow and now this*. Eventually some of them offered to help pack up Jenny's room, and Lacie told them exactly what she could do with their pity and offers of 'help'. Lacie's friends didn't come around for visits after that.

That night, she woke to the sound of buzzing. At first she thought that some of the bees had gotten inside, but, when she looked through her bedroom window, she saw small dark shapes gently landing on the screen and alighting again. There were also flashes of light, fading in and out like lightning bugs but the wrong color. Lacie had no idea what could glow that strange color of blue, but she knew it couldn't have been her hive. Bees don't phosphoresce, and they don't fly at night. There might be some stay-outs during honey flow, but not like this. It couldn't be her bees… but what if it was? It was the thought of her hive queen, lonely and missing her children, that propelled her up and out of bed.

She walked outside in her slippers and robe, flashlight in hand.

"Hello?" she called, feeling silly as she did so. She could still hear the buzzing. Lacie glanced at the shapes in the air then headed for the mud-room where her veil and gloves hung neatly from pegboard hooks.

Halfway there, a small blue light began to dance in front of her. Lacie stopped, and glanced over her shoulder. There were more lights there as well, but they were moving away, going around the house. No—that wasn't right. She looked closer. They weren't going around the house, they were going *into* it. Her flashlight beam played over the siding, then froze. The window to Jenny's room was open.

"No!" Lacie sprinted inside, protective gear forgotten. She streaked through the entryway and hall, her slippered feet patting on the carpet. At the door to the Jenny's room, she could already hear the buzzing. She burst through the door.

The bees were everywhere. They crawled on walls, over Jenny's dresser and nightstand. They covered the U of M poster that hung over Jenny's desk, their strange blue phosphorescence matching the "Go Blue!" slogan. Lacie froze, and the bees descended. They landed on her, not stinging, but gently touching her skin. She kept calm, partly to not alarm them, and partly because the light contact was soothing. They were on her clothes and in her hair. Her scalp buzzed, hundreds of tiny wings making her skin tingle with their movements. The bees flew out and away, strands of her hair clinging to their legs like pollen, pulling away from her head and creating a halo. Their abdomens lit up, not fading now but staying lit, a cloud of electric blue swirling around her that lit the darkness and hummed with a buzz loud enough to resonate in Lacie's chest.

As one, the bees leapt into the air. Swirling, they came together in a hovering, mobile mass the size of a large pillow. The swarm ball hovered in the center of the room just a few feet from Lacie's head. Its phosphorescent blue was soothing and strange all at once. There was movement and the swarm extended in some places, contracted in others. It began to pull tighter and took on a definite form. Lacie caught her breath. It was a face.

"Jenny?" Her voice cracked, and she tried again. "Baby, is that…"

The buzzing continued, and the swarm seemed to tilt ever so slightly, smiling at her. A crush of emotion took her breath away as she remembered the first time she'd seen Jenny's smile, a tiny girl in her arms, both of them swaddled in hospital linen.

Lacie stretched out a hand, not quite reaching the face in the swarm. "I miss you," she whispered.

Jenny's face moved, the bees pulling in and contracting. There was no voice, but Lacie could read the movement of the lips. "I miss you, too."

Lacie struggled for more words but nothing came out.

The bees shifted and Jenny's lips moved again. "I'm sorry," she said, "that I never said goodbye."

With the last syllable, the blue glow began to fade and Jenny's face grew indistinct. The swarm was dispersing. Panic rose up in Lacie's throat. She looked around, desperate to find something to keep the bees—to keep Jenny—just a little longer.

Strand by strand, Lacie's hair fell back around her as the bees left, each one flying alone into the dusky night. She realized that there was no keeping them; there is no keeping anyone when it's time for them to go. The buzz faded with the bees' departure. Soon, the only sound was the chirping of crickets and gentle, racking sobs as Lacie said goodbye to her daughter.

The Radish Field

Tarah Walsh

I was six years old the first time I picked vegetables on the farm. I walked the sandy dirt paths away from my grandparents' house into the fields with my mother and sister, following the tractor treads in the ground. The sun was hot on my neck but the air smelled sweet and the warm breeze kicked my hair in wild tendrils behind my head. My sister held my mother's hand, swinging it slightly as she stumbled over rocks and dirt clumps. We were going to pick radishes in the field.

I heard the buzz as the farmhouse disappeared behind the tree line. It was a soft grumble that grew louder and louder with each step toward the radishes, a constant hum that ebbed and flowed like a single being. My mother called to me, warning me to not scare the bees. The noise made my hairs prickle. I had never seen so many bees together at once. Never did I wonder where the solitary bee perching on my favorite lilac bush in our backyard went after I shooed it away. I knew now. Or at least I knew where my grandfather's bees went. I saw the white boxes teetering ahead; saw the black dots circling to and fro. My sister stumbled and grabbed my mother's hand tighter. Bees. Hundreds and hundreds of bees.

We passed the buzzing white boxes with ease. I stared at them from the tractor path, trying to make out a single outline of a bee. My mother pointed to the field ahead and we walked on, the bees vanishing behind the rise of the land. We plopped down on the ground and began tugging at the radish tops, pulling them out of the earth and brushing dirt clumps from their roots. "Rub them on your jeans," my mother said, cleaning one on her pants before taking a bite. We sat there in the field, my mother, my sister and I, tasting the tangy zip of radishes and relishing in the warm air of early summer. My sister chattered and my mother sang and all the while the bees hummed softly in the distance.

My grandfather told me that the bees helped the plants, made the vegetables on his farm grow big and strong and made the plant blossoms turn into tomatoes, cucumbers, watermelons and radishes. Older now, I no longer needed my mother to escort me in the fields and, as I eyed the boxes, my hairs no longer prickled at the sound. I watched a stray bee fly past me and I wondered at her fuzzy little body. I wondered where she was going—to the tomatoes or maybe to the wild raspberry patch beyond the tilled land. Her life was filled with work. Constantly buzzing from flower to flower, making the plants grow big and strong. I hummed quietly, mimicking their sound. I let the noise envelop me. It grew louder as I stood there in the tractor trail and I imagined the workers circling inside the hive, living their lives in the constant hum. After a while, I headed to the radish field still humming with the bees.

I was sixteen when I reached the white boxes and heard silence on my way to the radish field. I stood in the tractor trail across from the hives and tilted my head to the side. I heard the twittering of a nuthatch, the rustle of leaves, even the soft splash of ducks in the faraway pond, but not the humming of the bees. I squinted at the boxes, searching for the black dots buzzing through the sky.

I stepped off the path and inched closer to the boxes than I had ever been before. I strained my ears for the familiar sound. I crept closer until I was standing right over the crooked boxes and listened to the silence. The sun kissed my face and I felt a bead of sweat drip down my neck as I tucked a stray hair behind my ear. I turned back to the tractor path and followed it to the radishes. Once, I looked back, and saw the tops of the white boxes vanish behind the ridge. I tugged at a radish top and it popped from the dirt. I sat on the ground examining the fleshy red vegetable, its roots dangling earth behind it. Humming softly, I rubbed the dirt from the radish against my jeans and took a bite. Feeling my throat burn at the taste, I stopped humming and sat in the silence of the radish field.

The Honey of Oil

Rachel Channell

The honey of oil oozes over the tongue,
seeping into the senses with torpid
tenacity. It bleeds down combs
of iron and steel, a viscous tide
of ink-black industry.
The honey of sugar is gone, the ones
who made it outmoded by bionic bees
with wings of fiberglass and stingers
of silicone—mechanical creatures
who crawl through skeletal hives,
professing their progress with
a hollow metallic buzz.
I remember when the bees pulsed
with life instead of horsepower
and flowers bloomed in place of
the chalky plumes of smoke—
the honey of oil holds
no sweetness for me,
for in a world of automated
insects with gears and gas lines,
there is no need for beekeepers.

In Golden Slumber
Liza Mattison

It was August when the worker bee, leaving on a hunt for sweet nectar, sealed the fate for her entire hive. She collided with and got stuck in what would turn out to be Bates's right sock, and the poor thing panicked.

* * *

"Your father's sting is worse."

Bates, the man whose leg is being discussed, slumps against the arm of his wheelchair, inspecting a morsel of lunch he has excavated from his right molar. Despite the sunshine, a sepulchral gloom lingers about the room because of the towers of index cards. Stacks reaching to the ceiling cover every surface of the small room. They crowd the Oriental rug and spill across the pine floors. Smaller piles swamp end table and window-sills. Hundreds of thousands of index cards in all. On each one, a string of numbers written in Roger's cramped, deliberate hand.

"I've changed his dressing but he'll need his antibiotic this evening." Lovey says, though she knows Roger will not do it. Her love for Bates is as tender as a daughter's and feels acutely the injustice that Roger should be his sole legacy. People say Roger is some sort of genius, but to Lovey, Roger's intelligence helps no one, and this makes him an idiot.

* * *

The first thing Lovey did was call exterminators. Flushed out, the bees started anew in an old tree several miles away as the crow flies. Or in this case, as the bee swarms. The second thing Lovey did was declare the outdoors permanently off limits. Bates would never be stung again. Not on her watch.

"Want to work on your puzzle?"

Bates shakes his head. He is afraid to fail in front of Roger. Roger clicks two pieces together. There. His tone is bright. It is also condescending. Roger snaps two more bits together. Bates is amazed by the ease of his expertise. Then he realizes he too sees a match. A small victory. He sets himself to the task. Puzzles are frustrating more than pleasurable, but he doesn't mind. Frustration doesn't unnerve Bates. It's futility that terrifies him.

Roger, satisfied, returns to his index cards.

***

A Drone honeybee's pursuit in life is to mate but the force of insemination is so powerful it rips him apart, ensuring that what is arguably his first undertaking in life will also be his last.

***

Roger's pursuit is simple in its aim: he wants to conquer phi. Phi is a mathematical proportion that has variously been called the golden ratio, the golden mean, and the golden proportion, and ever since Pythagoras and Aristotle first wept at its exquisite majesty, men have sought to capture phi. The Greeks preserved its beauty within the walls and pillars of their Parthenon. Da Vinci sealed it into *The Last Supper*. Artists, engineers, and philosophers have all beheld its golden magic, and hungered to wrest it for their own golden glory.

***

Drone honeybees emerge from an unfertilized egg, and so have only one parent. The female who produced that egg has two. This means the number of ancestors in each prior generation for a Drone forms a Fibonacci series:

1(drone) ► 1(parent) ► 2(grandparents) ► 3(great grandparents) ► 5(great-great-grand-parents) ► 8(great-great-great grandparents) ► 13(great-great-great-great-grandparents) and so on. Successive ratios in a Fibonacci series (1/1, 2/1, 3/2, 5/3, 8/5, 13/8 etc.) get progressively closer to the golden ratio but can never equal what stretches without end or repetition, forever.

1.618 033 988 749 894 848 204 586 834 365 638 117 720 309 179 805 762
862 135 448 622 705 260 462 81…

***

Roger will saturate every cell of his brain with Phi's infinite, irrational
numerical expanse. He has already replaced the taste of mushrooms,
how clutches work, the scratchy warmth of wool against skin, and the
word dungaree with sequences of phi. He can no longer tie his shoes,
button a sweater or recognize if he is cold. He has forgotten Shakespeare,
how to bid for trump, and that his mother smelled like lilacs and flour.
One day he will not comprehend who he is and he won't know that he
doesn't know. Nor will he care, for there will be no more curiosity, fear,
ambiguity or doubt. Indeed he will feel no emotional sensation at all.
Roger finds great comfort in this, the beauty of nothingness.

***

*If a lesser wax moth slips past watchful eyes of worker bees long enough to
lay eggs, her emerging progeny will dine in splendor on honey and beeswax
and pollen, and also the empty shells of expired bees. Any animosity the
bees harbor towards the insatiable appetites of their unwelcome guests is
surely tempered by a grudging gratitude for the moths' disposal of apian
carcasses that would otherwise clutter the hive.*

***

Bates knows he used to be a man of importance. More than that, he
was a man of purpose. When he started, he possessed only a work ethic
and sense of honor. Those served him faithfully. The young new presi-
dent said, Visit anytime! You're always welcome here! But Bates declined
because a startling and awful truth had opened up like a terrible void
before him. And he embarked on retirement feeling as useless as a tooth-
less rake.

***

*Wax moths descended upon the abandoned hive and feasted with impu-
nity. Freed from its comb, the honey moved sluggishly for weeks in the cool
autumn temperatures. Until the sun sprung it from its torpor.*

***

At five, the room is faintly darker. Roger dimly recalls there is something that he was supposed to do but doesn't remember what. A dark, domed drop splashes on his card and begins to spread. Roger looks up. A crack in the ceiling hosts a row of small droplets. One stretches and with a viscous snap, plummets onto his wrist. He tastes it, but except for tomato soup and water, he has replaced every flavor with digits of phi. The drop tastes like neither tomato soup nor water. Another drops lands on Roger's shoe. Roger returns to his cards.

Bates is dozing. The puzzle pieces have mysteriously become sticky and there is nothing else to do. Roger chants numbers. To the sleeping Bates, the sound recalls the ambient hum of belts and gears spinning in the mill weave room. When Bates was overseer, he would savor those precious moments between shifts and let the leathery whispers of industry wash through his mind like a warm tide.

Drops rain upon his head and arms. They make lethargic trails down the bridge of his nose, get caught up in the fur of his eyebrows. He wakes and dabs a finger. Sweet mingles with hints of shampoo and salty sweat. Roger continues adding digits to his brain. Bates has never understood this inexplicable ambition of his son's. But Roger was always a stranger. It became habit to think of Roger with a mixture of disappointment and regret and so he learned to not think about him. For sixty years each has existed around the other. Separate worlds orbiting around loneliness.

Honey pours in profusion now. Its weight strains a ceiling panel. Soon a cascading sheet of amber spills down into the room. A film slides down Bates's face, coats his skin but he has fallen asleep again. It seeps over Roger, who recites digits, his voice never faltering as liquid oozes, wrapping him ever tighter in amber arms. Stalactites drip into Bates's open mouth. It's hard to breathe. In dreams he can't escape. Roger recites. The numbers march. To Bates they become the thrum of industrial looms but have grown loud and menacing. They become a demanding alarm; now the clack of trains, trucks, delivery men; now the wail of a newborn son. Sounds all around him, scream, swirl through him without beginning, without end. A drowning fog. Unmoored, he gropes but can see no horizon, no limit, no escape. He is lost. Lost. Lost.

Honey oozes, creeps down Roger's shoulders and head. Inching amber beads trace contours of his body, fill the hollows between his knuckles.

The divots of his shoulders. He keeps adding digits. Honey stings his eyes. He doesn't look up, doesn't notice the cramped landscape, with its towers of cards covered with phi, now glistens gold.

***

But a hive is a finite thing. It does not continue forever and eventually the honey stops. Long after Lovey has moved on, left the Bates home far behind her, she will still remember her arrival that following morning. Her memory won't accurately recall the panic that propelled her into action after a seeming eternity, or the fearful dread that she was staring at the first dead people she had ever seen before. What will remain, when all the other details of that day have faded and blended together and been distorted by memory and exaggeration and years of retelling, will be the startling and almost breathtaking beauty of the scene she stumbled onto. A golden landscape, with its two inhabitants, like creatures from some forgotten world, preserved in peaceful slumber under eternal amber shroud.

The Land of Milk and Honey

Alixandra Nicole Prybyla

The honeybee is economic. Who knows Mother Nature's currency better than her tax collector? Heralded by the loud jingle of gold in her pollen pockets, the honeybee campaigns in a meticulous system. Her wings ring like the town crier, "I'm coming for my dues, pay up!" When she returns to her hexagonal Wall Street, she is an Angel Investor; she jinks, jukes, and cajoles her way through the blood sisterhood, all operating as brides to their business: the hive, their chief operating officer. When Moses, glaciated in fear, thawed his piety at the burning bush, he was guaranteed acreage flowing with milk and honey. Little did he know, at the end of the pause was an assumed clause: bees as his national bankers; his mentors.

Shalom Lilker proclaims that the Holy Land's kibbutzim—specialized socialist communities that freckle Israel's sunny face—must have the acceptance of self-discipline from a source without themselves, as a group, not within. Structure. It used to be that children in kibbutzim were taken and raised and schooled separately from their birth parents to avoid preferential treatment. The community, geometrically fitted, slumbered in a single room. Family before the self. An acquaintance of mine took flight from New York one summer and worked on a kibbutz. The hum of men and women buzzing about their daily lives on Sde Eliyahu, he remembers fondly, contributing to the collective honeypot from which they all dip into, crescendos pleasantly as they craft bee boxes. Perhaps, I think, this kibbutz is the ultimate form of method acting. So like bees are humans. So much we could learn.

Now in Israel, I oscillate between the Northern and Southern villages in far fewer than forty days and nights, at any given time only a few hours from each border; I peer out and over into Lebanon, Syria, Egypt, and Jordan as my bus trundles along roads like pebbled skeins. We scaled Mount Bental that misty evening.

At the top, the Golan Heights open up like a dark velvet purse. The horizon gulps the oyster sun, spits back pearls to spin around the black basin above. It was all as Ovid said, humans, purple clay sucking at our ankles, faces to the sky; the great basin inscribed with stars that, when trodden about in puddles, disperse with a watery whisper. This is where we came from, the veins of the rocks becoming our veins, the hard parts becoming our hearts, maybe. Every city is pestled down to pastoral green. No urban sprawl. It is the perfect place for an organic miracle, all that gangly shrubbery, all those Israel-Syrian fences tinseled with barbs that crown the hill's crown like a thorn crown.

"Last year we saw bombs being dropped right over there," one guide dwarfed a distant Syrian town with a sweep of his hand. We stood as a group with our backs to a captured war fort wrung up in chicken wire and stone blocks. "There's the United Nations—see, nestled between the hills." If not for the government-occupied space, I wouldn't have been able to see where Israel bowed into Syria. But the borderland is ribboned with evident strife that goes back much farther than their defining six day war in 1967. The wind blows for us held above both planes.

Ohav is an Israeli soldier, but in that moment, bent over the railing that secures Mount Bental and offering me a stick of licorice gum, she is another young adult, eyes wandering, mouth working through candy and a song: "Erets zavat chalav ud'vash. Erets zavat chalav ud'vash." A land flowing with milk and honey, she sings under her breath, to me.

'Milk and honey' had never made sense to me before that moment; maybe it's the bee-boxes just down the mountain that catch my eye, propped up on flat tires. Some workers dare the ascent in search of flower business, and I can see their nectar bellies fluoresce amber-yellow in the last yawn of light. Some come to rest, hidden in the arms of highland petals. When they dare to buzz too close to us they risk their lives to misunderstanding. What do bees think, I wonder, when they approach us as cautious natives, cupping the day's spoils? Their buzzing must be an inquiry about our fees to the land. I'm sure of it, because being a Land of Milk and Honey means advancement: it means cultivation to a point where we can sustain more than ourselves, and reap those benefits; more than day to day, we can be far-seeing—the honey process is the vessel in which time comes to mature. The bees are clever entrepreneurs: their business plan is extensive, comprehensive, and inclusive of all in their world's radius.

On either side of me is a remarkable stretch of green with its own bee-hives. Their occupants barter amongst each other. They zip above barbed wire fences, below, slip between unnoticed by border guards and leery-eyed neighbors. I would like to remember how to live together as the bee does. It would be a grand existence, a mature, fluid, respectable existence, to live so close as individuals that you cannot part, to hinge yourself to your hive and float into advancement amongst your cousins as if it was the most natural necessity in the world. I think again of the kibbutz in its mimesis, and wonder.

WITH THE SEASON

The Raven and the Bee

Sarah Herdan

Long ago there lived Raven and Bee. The two had become friends when they happened to land on the same branch of an apple tree. In late fall, the two would fly back to the old apple tree and perch together. Bee: elegant, royal, and of few words. Raven: a tremendous, feathered, shadow of a creature.

One crisp fall day, Bee spoke to Raven while she hovered at his eyes. "Why don't you watch over the world in the winter while the bees restore and rest in the hive, and I will watch over the world in the summer while the ravens hide from the heat in the branches of the Hemlock trees?"

Raven stretched out a wing for Bee to rest while *he* spoke. "Yes, let each of us mind the world for a while and then we'll come back to speak again on this tree." Bee looked away toward home. "It will be a long time, Raven, before I sit on your black wing again. Goodbye, Raven."

***

For many winters, Raven flies over the land. White snow and ice fall like heavy lace on the trees and the wind whips the mountainsides. He sees men and women carrying water to the animals. Horses walk in fields with their bodies wrapped in warm blankets. He watches while the women sit in circles around the fire and make bowls for the villages.

Raven dies and another is there to take his place. Like a great bodhisattva, when Raven dies, another who is both the same and different is there to take his place. And so it goes.

The world orbits the sun a thousand times—a thousand years have passed. Raven continues his watch over the winter skies. He flies down the road and lands square on a weather vane so that he may see in all directions. He looks at the cars moving by, sees fields of barbed wire, watches water flow down through ice in the brooks, and always he calls out to the empty rose vines hoping Bee will be there to hear his winter "hello" the following spring.

The seasons turn, turn, turn. Every year, winter hands his torch to spring.

In early spring when the snow looks like icy corn kernels, the frost finally begins to melt. Hives wake up with a great thirst. Bees fly out for water first and food second. Passing through a tiny circular tunnel door, the bees search the Earth for water by scanning brooks, ponds, and puddles. Once they have quenched their thirst, the bees fly down into the daffodils and lilacs. The flowers smile up at the bees as they make their steady descent into them. Each offers the other life. Bees fly through fields flooded with pollen—each worker bee looks as if he has been dipped in powdered gold. One bee is every bee. The Queen's mission is the mission of all. Bees hear no word other than the word of the Queen. The clock ticks forward. The world orbits the sun a thousand times and we have now caught up to the Raven. The bees still eat; however, the honey does not taste as it once did.

The flowers still bloom, but the smile they once shared freely with the bees has all but disappeared. Flowers have been poisoned by the dust of man, but when your only food is poison---that is the food you bring home to the hive.

***

Raven and Bee meet at the old apple tree. "Here we are again," calls Raven. "I watched you bees through the branches of the trees every year, but no bee ever looked in my direction until now. It fills my heart to be here with you again. Tell me of your summers while I took my rest in the branches of the Hemlock trees.

Bee turns towards him, lands, and rests on his wing. "Well, Raven, I had never asked them to look up from the flowers to see you in the trees. I heard your 'hello' on the rose bushes year after year. I left my 'hello' for you between the stems so that it would get caught in the thorns and reach your ears in the heart of winter. We eat men's creations and inventions. The choices of man now mingle with the pollen in our flowers. Now my bees breathe a sigh of relief each Spring when the flowers return as they should. Many of us have died, but all is not lost, Raven. Here I am to sit with you. "It has been such a long time since we sat together on this branch.

Tell me of your winters, Raven. Raven looks at Bee, "The mountains are still covered with ice. I have found the same taste you speak of in the brook water.

I have flown where it is so cold that the flowers do not bloom. Metal, brick, and wood boxes stand in the places where round houses once stood. I watch over the world now and I must peer through the haze of man in order to see it. Don't you miss the words they once spoke in our direction?" Bee nods and waits for Raven to continue. "I seek the places in this world where the trees still stand, the flowers still bloom, and where our voices are not silent. Let us continue our watch.

The Queen of the Bees sits for a moment and gathers her courage. "We agree to continue. When will we meet again, Raven?" Bee looks in the direction of home. With steadfast faith and courage the King of the Ravens calls out to Bee. "We will meet again when the circle is no longer broken and when the flowers, once again, smile up at you as you fly down into them.

Hospitaliano
Daniel Talamantes

Jose and I were about a half-an-hour into the thick of Yolo County before we pulled into an almond orchard. It was my first time out in the field, even more, it was the first time I wore the apiarist uniform.

Before I hopped into the truck back at the ranch, the owner said, "As of lately, every time we hand out these uniforms, the person manages to quit within the next week."

I took the uniform gladly not really considering what he said until we started driving. Instantly, I felt guilty, because driving through acres of sunflower fields, those words incepted the sudden burden of obligation.

We drove awhile more down the dirt path after turning off the main road. I looked back at the stacked rows of sugar-filled canisters in the trunk to check if they were stable. They were strapped down by a few industrial chords, but my idea of physics wanted to betray their integrity.

Jose told me not to worry. He was maybe a year or two younger than I was, which must have meant he was around twenty or so. It was remarkable to me that the owner of a multi-million dollar company could trust a twenty-year-old with not only driving a large truck but, hundreds of dollars worth of liquefied sugar in the trunk. Nonetheless, he seemed like a sound guy.

To the left of our path was an aqueduct that bordered an almond grove and to the right was a tilled field. In the middle of the articulated mounds of sodden dirt, a solo man was reconstituting a blocked passage of water. Even from our distance I could see his skin was leathered. He wore a wide-brimmed straw hat and denim over-alls that revealed his impossible muscle mass. He was almost like another breed of man, born of stone, as he stood there digging away at the fertile soil, laboring to make nature work in his favor. I looked back at my skinny, unblemished hands and frowned.

"So, are you family?" Jose asked.

"Well, my partner is the niece of Jerry," I explain—Jerry being the owner of the Apiary.

"I see. So how long you been working now?"

"Well, I don't know if I'm technically an employee or just helping out while I am still in town."

"Okay. How you liking it?"

"I actually like it a lot. Much better than my other job anyways."

"What else you do?"

"I'm bussing at Olive Garden."

He laughed, "What's that slogan again?"

"When you're here, you're family."

We pulled into an entrance in the almond trees where a hundreds of hives occupied a good portion of the glade. Their was a thin atmosphere of bees moving about the place. They came crashing into the windows and began chewing away at the excess sugars of the canisters. We zipped up our suits and fastened our masks in the safety of the truck.

We then entered the fury, bees instantly landing on the threaded cage of the mask, as if prisoners gripping onto the bars of the jail cell. It was a hail storm of bees that took a moment to adjust too, but after a few minutes it was all waterfalls. Jose grabbed the fork-lift tucked away in the trees, unloaded the first pallet of canisters, and drove it towards the nearest portion of bee hives.

He got out of the fork-lift and motioned for me to watch. Grabbing two sugar canisters, he leveled one over a cut-out circle on the roof of the hive, flipped it quickly, fixing the lid of canister into the slot, and then did the same to another.

Looking at me he to see if I understood the procedure, I gave him a thumbs up with my rubber gloves. He then looked down and yelled something I couldn't hear over the buzzing din.

"What?"

"Are those your shoes?" he asked again. I looked down at the Nikes, ruined with paint, dirt, and blackened sugar, and looked at his laceless boots. I shrugged and he laughed. I wasn't quite sure how to answer his rhetorical question without looking like more of an idiot. He patted me on the shoulder.

The feeding went on for nearly two hours: placing canisters of sugar on about a hundred hives. The heat began to take a hold of me. It was June in the central valley. I never asked but you could guarantee the temperature was flirting with the hundreds. Inside a plastic suit and rubber gloves it makes conditions even more dramatic.

Half-way through we take a break near the truck. Under the bed is a tank of water, of which he pours out into a bottle. He squirts the water through the mesh of his mask then hands it to me. I just then realized that it probably wasn't the best idea to take the mask off so I sprayed the water through too. Most of it ended up slithering down my chest in uncomfortable ways.

"Not so bad, huh?" Jose asks.

I look up at him. "No, not really bad at all." And I meant it. It was certainly difficult and tiring but there was a redeeming, fulfilling quality about it.

In the last hour I nearly, either from an elevated state of labor or heat exhaustion, transported into a somnambulist state. Before placing the canisters on the hives I'd watch bees tearing at the limbs of imposters with a reverence. Everything about the bees' behaviors bewildered and charmed me, even their most appalling, cannibalistic acts seemed religious.

Something about the infinite regression involved in bee-keeping philosophically hit the point. The way it deconstructed from international market trade of almonds to an almond grove in Davis to thousands of bees pollinating the orchards to a hundred beehives in the glade of an almond orchard to a few hundred bees in each beehive to a queen in each hive. Each piece working to make the other happen, all important, not individually, but as a whole, in a utilitarian ideal.

Driving back in silence, I watched the fields of sunflowers swaying to the gentle wind. I took off my rubber gloves to find my hands completely white from the lack of oxygen. Jose saw me staring at them and laughed.

"Good work today," he said.

Back at the ranch he dropped me off at my truck. I looked around to see if anyone was around but it was empty, and besides, the sun was setting now, it was time to go.

***

I threw the bee-suit in the bed of my truck, drove twenty minutes to the parking lot of a suburban shopping mall and parked in the employee lot for Olive Garden. Hastily, I put on my work pants and collared shirt and walked in through the entrance.

The hostess at the front gave me a fragile greeting as I passed her into the kitchen. Ignored by nearly everyone else I walked to the bathroom to wash my face and hands. While soaping I noticed a roughness about my skin and smiled.

As I walked out the manager confronted me and asked, "You check in?"

"Yeah."

"I need you to grab drinks for table five. Busy night, really need you to be a…a busy bee. You should know something about that, shouldn't you?"

"What drinks?"

"I don't know! Ask the waiter!"

I flipped her off as she hustles away.

I asked for the waiter and finally found her. "I already did it. You were too slow. I'm really gonna need help tonight, your pace isn't going to cut it. I have two tables!"

"Two tables, huh?" I asked, tasting the sarcasm feeding a grin.

"You know what…I'll get help from someone else."

After she stomped off I grabbed a tray and walked over to clear a table. All the plates had an excess of food that was going to be thrown away. There were wines unfinished and chocolate deserts never touched.

After wiping the table I studied the sham interpretation of Tuscan architecture in the restaurant. I read my name tag, under my name was the word *hospitaliano*. *Hospitaliano*: hospitality and Italian-*o*. Two things remarkably lacking here.

"Excuse me, sir. This chicken is cold," exclaimed a lady with died red hair and a purple coat. Her husband, with a combed grey hair, nefarious mustache, and grey striped-pin suit, eyed me over like I was the one who committed such barbarity onto them.

I didn't say anything because I was absorbed by the replica painting of an orchard framed above them.

"Hello? Pal? Her damn chicken is cold!" asserted the man.

"What is going on here?" asked the manager as if accusing me.

I look at her, then the two below, then the other customers and waiters around who were eager for an answer.

"Well?" the manager bullies.

I didn't answer, I just stood there, gazing at the replica.

Lonnie's Front Porch

Deborah Darling

We cannot pass his house on 5th Street
without stopping to chat.
Lonnie has more for us than honey—
lined up in neat rows on a table,
golden pints and quarts in canning jars.
He warms our journey
down this broken sidewalk
with a flow of amber conversation.
The rooms of his house, unattended,
lie cluttered with old books
on the science of beekeeping,
cold cups of tea and stale crackers.
My daughter lives next door
and knows it won't be long
until the porch is empty.
Honey can't cure cancer.
She brings the children over
with hugs and cookies,
buys a jar of raw honey
and prays
for a few more days
of sweetness.

Letters After Achilles

Stefanie Brook Trout

I.

May 2, 2013

Dear Bees,

The cold air settled on my exposed back this morning, and I awoke from the chill. Snow shrouded the world outside my window, colonizing the tree branches like arctic lichen.

Yesterday, they warned us this storm was coming. They termed this record-breaking storm Achilles. I didn't believe them. Just two days ago, I donned a tank top and shorts, baring my pale skin to the bright sun and sweating in the eighty-degree afternoon. This morning, it's as if spring never arrived.

From the window, the snow looks soft and quiet just as all snow appears from afar. Outside, however, the snow is neither soft nor quiet. The temperature hovers just above freezing so the snow does not drift as gentle flakes but rather drops as airborne slush, pockmarking the white carpet upon impact. The fat white tears cannonball into puddles on the street.

It hasn't been a week since we in the recently-formed Bluff Creek Bee Club released you into your new hives at the Casey Land. If we had anticipated the arrival of Achilles, we would have waited for the storm to pass. But by the time the storm was foretold, it was too late. If your colonies weren't so new, I wouldn't be so concerned. If you survive this storm and the summer that must eventually follow it, you will winter out-of-doors and by then, hopefully, you will be strong. A superorganism adapted for life on all continents save Antarctica, your species has survived cruel weather before. But right now, your queens are fresh out of the cage. And I worry.

I want to save you from Achilles, but leaving you alone is the only way I can help right now. I hope you huddle close in your hives the next few days, ranks closed around your respective queens, shivering to raise your body temperatures. I hope you conserved your sugar water wisely, stretching the reserve until Achilles has gone and the air is safe for us to open your hives again for a top-off.

Since you're tucked away in your hives, you likely haven't had a look at the world outside. It's strangely beautiful. I hadn't realized how green and vibrant the grass had become until it was silhouetted against bright white slush. As courageous and ambitious as any Trojan, the grass fights back against the spring snowstorm, radiating the energy of life and thawing the dimpled snow before it even stops falling.

I promise to write again soon.

Yours,

Stef

II.

May 4, 2013

Dear Tallgrass Prairie,

The snow melts in the spring sun, dissolving into dewy grass. Good news for bees.

The bees' home at the Everett Casey Nature Center and Reserve is seventy-six acres straddling Bluff Creek. Five years ago, 1946 Iowa State Engineering alum Everett Casey gifted the land, valued at $201,000, to the Master of Fine Arts program in Creative Writing and Environment. Why he did this, I'll never know for sure. Casey cited an excellent writing class that he took at ISU for an explanation. A single class. I like to think that it was for me and others like me in the MFA program. Casey was from Detroit—not too far away from my own birthplace and childhood homes in West Michigan. I like to think that Mr. Casey knew about my type, that we would need a piece of wildness to be able to make Iowa home. I know it sounds selfish, but I like to think he reserved this place for me and others like me so that we never forget the privilege we have to live on this land.

The property should be carpeted with tall prairie grass, as should the majority of Iowa. Forty percent of the United States was once covered in you, my dear Tallgrass Prairie, but Iowa led the rest of the union with the largest percentage of its land area devoted to the native grasses—a sea of grass that can be just as disorientingly awe-inspiring as the open ocean or, something I'm more familiar with, a Great Lake. Your beauty has always been subtler than that of other landscapes. You hid much of your treasure underground in your amazing, complex root system that held the wet, rich soils in place.

Now Iowa leads the race to the bottom—with more than 99.9% of its natural landscape gone, replaced by a system governed by drainage tiles and an excessive amount of chemicals. Thirty million acres of big and little bluestem, Indiangrass, and switchgrass—all plowed under to make room for cornfields. Eight thousand years of prairie legacy disappeared so Americans can get fat on soft drinks and corn-fed cattle. I prefer to get fat on honey.

I miss you.

Love,

Stef

III.

May 10, 2013

Dear Casey Land,

At last, the skies are calm enough to expose the hives to the elements. Emerging apiarists pile into a van and set out to visit you. We must check on our bees—make sure the queens are still alive and the workers are building comb for brood and replenish the sugar water. Soon the bees will feed themselves.

We burn newspaper in the smoker, pumping the bellows to fuel the fire. Standing to the side of the first hive, we puff smoke into the openings. We lift the outer cover, give them more smoke, and then remove the inner cover to reveal the built-up frames. Right now, the hives are short with only the lower deep in place. The lower deep is the brood chamber, where the queen lays her eggs—as many as two thousand per day. Soon we will add the upper deep, the food chamber, and a few weeks after that, a shallow honey super. We don't expect to harvest much honey this year, but we hope we'll each get a taste, at least. But for now, we just hope our colonies survive.

We pull out frames for inspection and find our girls have been busy despite the chill. They built extra comb between the frames, too much comb in places, and a large piece of the hexagonal wax breaks off and falls into the hive. A brave beekeeper sticks her gloveless hand into the depths of the chamber and pulls out the fallen chunk, covered on all sides with bees. After gently brushing the bees off and back into the hive, we save the piece as a souvenir. Later, on the drive back to Ames, we'll notice the tiny rice-like eggs the queen has laid in each cell, our proof that Her Highness is performing her royal duties.

We shift some frames around, refill the sugar water, close up the hive, and then do the same with the second hive.

The business of our visit taken care of, we decide to enter your lovely forest to look for blooming wildflowers. We hike down the steep hill, through invading cedars and into the oak forest, and past the agricultural field recently converted from corn rows to oats in preparation for alfalfa planting—your only source of income and now on its way toward becoming a much better source of nutrition for the bees. At Bluff Creek, the fast water fills the channel more completely than we've ever seen, our feeble rock bridges submerged by the snowmelt, creating exciting riffles.

Bluff is a meandering creek, and we follow its winding way downstream to the sandy point bar that demarcates the edge of the property. Though our English Department holds your deed, we stewards don't like to think of *owning* you. You aren't *our* property. You've been home to all varieties of native and invasive flora and fauna, and now that includes a pair of honeybee colonies, but please don't get the wrong idea. This kind of colonialism is rather different than the kind you might be apprehensive about.

It's funny how beekeeping has caused us to break the only two rules of the Casey Land: take nothing and leave nothing. We left a couple of hives packed with tens of thousands of honeybees. For now, we've just taken the broken chunk of comb, but we're hoping to take honey home eventually. And beeswax. And propolis.

On our way back to the van, we lose ourselves in your woods. We accidentally follow a deer path forking off our main trail. We emerge from the woods on the edge of a freshly planted field and follow it toward the road hoping to see our hives just around the bend. We don't.

So we reenter the woods, no longer on any path at all, blazing our own trail that includes crawling under and climbing over fallen trees. We shinny down a gully, hop the muddy bottom, and scramble up the other side. We scale another barbed wire fence—a promising sign—and emerge from the woods again. This time, we see the white hives and know that we are home.

Warmest regards,

Stef

Patron Saint of Bees
K.E. Kuebler

I suppose the bees changed everything. Until then, the day started no differently from any other.

Josephine, my younger sister by thirteen months, groaned her protest from the next bed when I shoved back the curtains to a new morning. Our mother had traded her life for Josephine's in a difficult childbirth and our father, unwilling to muster the effort for a second courtship and marriage, grudgingly accepted that there would be no son. He commandeered that ball-fisted infant instead. He was Joseph. She was duly christened Josephine, called Joss by all who knew her.

Father and Joss are as alike as any man and his only son, both mean-tempered and obstinate, both busting out with all the charm and charisma of an evangelist in Indian territory. I, with my quiet ways, am continually dismissed by the pair of them as being "just like your mother," as though Joss herself had sprung, fully-grown, straight from our father's loins.

Maybe I do take after my mother. I have no reliable remembrance of her, merely a prick of loss on those rare occasions when my thoughts stray to the young woman she must have been.

I splashed my face with cold water from the basin and tugged on a split skirt and blouse and boots. Father was hunched over his coffee in the dining hall downstairs. He turned his head at my step, a bright expectancy in his face. He looked at me, then, and the light went out of him. I wasn't Joss. "Morning," he greeted me.

"Morning," I replied. I helped myself to strong coffee and buttered toast.

Stan Parks rose politely from his seat beside Father and I gave him a nod. Joss would have dropped into an exaggerated curtsy for Stan. It was a game they played. Joss played games with most men.

Stan Parks was a loan officer from the bank. Three years of drought were taking their bitter toll. Father would have to ship in costly hay from Missouri again this year to keep five thousand head of white-faced cattle alive through another barren winter.

I tucked the crusts from my toast into my pocket and pushed my chair back from the table. Stan swiftly rose once more from his own chair. My father ignored me.

I walked to the stables and Sourdough whinnied at my approach, though she couldn't have seen me yet, or heard me either. Sourdough was an improbably small mare, a numbing disappointment to Father given her carefully selected lineage. Had she been a suckling pig or a pup, she would have been judged a runt and likely destroyed. She was mine now, and I could thank Joss for it.

I believe Joss had perceived my secret attachment to the little horse for she dimpled one day with mischief and cajoled our father to "let the two misfits have one another." If her words were hurtful, surely they were unintentionally so, for those words had won me my precious mare from a man who could deny his younger daughter nothing.

Pedro, one of the legion of men in Father's employ, was waiting just inside the stables. He had saddled Josephine's horse and was leaning forward now over the gelding's neck. He heard me enter and whirled around toward me and some powerful emotion brought the blood into his face. When he realized it was only me, his visible disappointment was almost anger. I could have been watching my own father all over again.

Pedro straightened his shoulders, deliberately blocking my view of Joss' horse.

I stepped past him and managed to glimpse the wildflowers that he was braiding through the mane of my sister's favorite mount. Sourdough nickered at me from her stall. I palmed the bread crusts into her grasping mouth and bent my head to breathe against her flared nostrils while she chewed. I saddled her and any unhappiness I may have felt slipped away as she and I fled across the plains.

The land of my earliest memory rose up to greet me. Even burnt now to cracked clay beneath the pitiless sun, it was beautiful to me. Sourdough slowed at last to a walk and our passage parted a caramel-colored sea of pedigreed cattle. Deprivation had diminished the beasts to sharp bone and hide, and dozens of them followed behind us a half-mile or more, bawling at me for feed. I had none to give them, no way to ease their suffering.

I'd learned that ranchers south of the border were crossbreeding the white-face with the native longhorn, producing a cow that was heartier, better able to thrive on poor forage. I privately admired such ingenuity in the face of ongoing drought, but Father's vision of himself as a cattleman was too closely tied to the pure-bloodedness of his white-faced stock for him to change. Nor would he consider selling off part of his immense herd, not even to reduce competition among the starving animals for the scarce vegetation.

A wild hare broke cover on my right, and a hawk in pursuit dropped like a screaming arrow from the sky. The hare suddenly veered, darting to safety between Sourdough's hooves. Sourdough snorted her surprise but didn't spook. The frustrated hawk thrust up over my head with a taut snap of wing feathers and the talons, still extended for the kill, narrowly missed my eyes.

Sourdough would have drifted northward as we used to do, making for a plateau from which I could survey the northernmost boundary of Father's vast spread. Several days' riding are required to reach the ranch's end in any other direction. But I disliked the view from the north now with a new line of oil wells pumping along that horizon like giant buzzards pecking at the ground.

Oil companies were constantly pestering Father for drilling rights on his property and he had so far refused them. Persistent rumors of poisoned water holes and petroleum-slicked grazeland warned him off.

But oil promised riches. And I suspected that Father had mortgaged the ranch to the hilt. I'd overheard Stan Parks urging Father this morning to sign with an oil company soon if he hoped to keep current on his interest payments. I decided not to think about it now and turned Sourdough into the south wind.

We picked our way through blooming sagebrush and I became aware of a humming sound all around us. Honeybees, hordes of them, were feeding on the purple blossoms. They lifted like a cloud above us, then settled once more in our wake. Ahead of us, the bees streamed toward a stand of dead and dying trees. I guessed there was a hive there, and we trailed after them.

The bees were flying in and out of a mesquite trunk, split by lightning. I smelled the slow, sweet syrup of their honeycomb inside it. Father would be pleased. He relished fresh honey on his corn cakes, and I'd be the one who discovered it for him.

I don't know why I didn't ride immediately home with my news of honey for the harvesting. I meant to. Instead, I left Sourdough a short distance away and approached the hive on foot.

I stretched out my arm like it belonged to someone else and pressed my palm against that broken tree. Possessive bees burst forth and clustered across the back of my hand. They crawled up my wrist under my blouse and lit on my throat and face, but they didn't sting me and I wasn't afraid.

Their hive went deep beneath the soil. I sensed the reverberation of their humming multitude through the soles of my boots. It was as if the bees spoke to me and theirs was the voice of the land itself, and I thought I heard my mother's soft voice, too. They spoke together of Earth and her creatures and of my responsibility.

And I understood them. There must be no drilling on this land. I'd willingly lay down my own life to prevent it. Nor should the animals here be made to suffer any longer. Father's white-faced herd must be thinned out, at least half sold off to raise his necessary cash. We'd crossbreed the herd's remainder with the longhorn.

I wondered if I dared defy my father. If I did, though, what possible good could come of it? I was the daughter he despised.

No, much better to win Josephine over to my side first. Joss would be fairly easily swayed for she took no genuine interest in the ranch. I need only appeal to her vanity by making a confidential game of it, by challenging her to prove she could bend Father to her will. She would compel him, then, to do the bees' bidding.

I was loved. I knew it now. This land loved me, and this would be my strength.

Colony Collapse Disorder

Amy Newday

The honeybees are leaving, abandoning
their white subdivisions, their elegant
egg-plump queens, their wingless
brood sealed in its cells. The honeybees
are leaving and they are not returning
to their pre-fab honeycomb
condos, their pantries filled with bee
bread. Sugar-syrup delivered to their door sills
does not entice them. The bees are tired
of commuting cross-country in semi-trailers, the
endless almond blossoms, and then the cherries, row
after row. The incessant droning
of the tractors makes them mad. They are sick
of our shrieking, our sticky
picnics, the smoke from our barbecues, our poisonous
grass. The bees' wings
are folding like torn
paper fans. They are falling to earth, quiet
as angels. Everywhere, the sound of no bees
dancing. Tulips swing
in the wind,
empty as unclappered bells.

Brood Chamber

Daniel Talamantes

Off of Highway 50 from Sacramento, I take the overpass on through the north side of Davis's residential district and turn left on a country road that leads me into the fields of Yolo County. To the left is the first entrance to the Pacific ridge that will pass through undulating geographies full of wineries, expensive hotels, ornate mansions, and from there you'll eventually reach the ocean. From the road I'm on, every other point of view you'll find agricultural fields and orchards of sunflowers, wheat tules, corn, almond, chestnuts, what-have-you, extending as far as the horizon. It is a sea of bucolic zest with archipelagos of homesteads barricaded in walls of eucalyptus trees. And then there is that valley heat, that infamous heat that will one day melt this all back into a wax.

Anxious, I pull into the apiary at seven in the morning. With a whole company of emotions, indecisions, and thoughts raging through my head I walk out to the main drag where a whole assembly of suited employees are unloading freights of beehives from the back of trucks that had just arrived from Modesto.

Above them an estuary of herons, who found perennial residency in that batch of eucalyptus trees after they were displaced from the landfill some years ago, were vocalizing their interpretations of the activities below them. Those trees were planted there as windbreakers, but I guess over time they inherited their own form of distraction in those birds.

Walking toward the barn, navigating around the melee of upset bees, tired workers and condescending birds, I pass through the large ingress where more workers, with their suits stripped off, are cutting, stapling and painting new foundations for the hives. They greet me with their usual candor. Never are friendly insults in short supply when you're working with three languages and all their colorful dialects.

I pass through the back of the barn into a narrower, dim hall. Lined up in a row are a hundred or so canisters on pallets, and the owner's son is filling each one with liquefied sugar.

"James, do you mind helping me out here?" he asks.

"Uh…" I consider a moment. My aim was to talk to the owner right away, but the better part of me decided to help. "Sure, yeah."

I grab the monkey wrench, roll up my sleeves and start fixing the caps back on the gelatinous, stubborn lids.

"James, we really appreciated the help around here. Other college students we've had before haven't worked out so well," he says over the hum of the spewing sugar tank.

"Well, it's a great job. I like it a lot better than customer service."

"Well, I'm sorry to hear you are leaving. But I understand. What is your next move?"

"I don't know. I get anxious thinking about it."

"Well, I didn't start 'til thirty. You got time."

We finished up the last of the canisters. He walks outside to start loading them onto the truck, which will be delivered to various locations throughout northwest California.

Down the corridor a door opens and a golden light emanates from a small room. Inside the owner is sitting on a desk chair and looking through a large magnifying glass at the brood chamber mounted on the bottom board before him. He is in a white suit, his white hair in the manic state of a mad scientist, and glasses press tight to his nose.

"James, come in," he says.

I enter the small, golden room. He stands up and gestures for me to look. Inside the magnifying glass there are small combs filled with honey.

"Look closer, you see any larva in there?"

Inside one of the combs an almost undetectable string wiggles. "Yeah," I respond.

"Everything you see around here starts with that. This whole agricultural economy begins with that tiny creature."

I watch the little string vibrate in the jelly. Everything I want say, all the thoughts I've been mulling over, and all the uncertainties I've encountered over the last few years, halted at the sight of the larva. I felt I was watching the universe in the smallest thing.

"You started this way. Fields start this way. Ideas start this way. The universe starts this way," he continues. He is always one for overdoing the poetic in things.

I lean back. "Thanks for showing me that," I say.

"We appreciated you working here. You sure you want to leave?"

"Yeah. I think it's time."

"What are you going to do?"

"Don't know. Start new." I shake his hand and nod. "Thank you."

I stand there looking down through the room, to the corridor, then out through the entrance of the barn and start walking forward.

I shake hands with a few of the workers as I exit and listen one last time to the laughing society of herons in the trees. Driving through fields of sunflowers, I roll down the window, letting the summer heat melt me away.

Once I reach the overpass onto Highway 50, I take a right onto the Pacific ridge. Before me are wineries, expensive hotels and ornate mansions. I'm driving past those and heading for the ocean. From there, I don't know where I'll go. But I got time.

The valley is behind me now, and it is thriving without me. Once I reach the ocean, I'll start from the end. I'll look back on where I've been. I figure I can start new from there.

Building Defenses
Erica Eastick

He didn't even bother to knock. I found the letter on the porch, snug in the cushion my mother had made me before she died as if the bad news needed a rest in the old rocking chair before presenting me with its impossible demands.

Payment is way overdue, I know, but surely he'll give me more time. Mother's Day is coming up; the native bouquets usually sell well; the hives will come good now the weather's calmed, and the morning sun shines more reliably than it has in two weeks.

The wind has dropped another log. It'll need soothing and plugging so the stingless bees won't desert or die from invasion. Taffy's interest is held by something deeper in the bush, rustling and scrambling in the bracken, but too sneaky to be seen. The black and white brush of her tail wags lazily, the hunt a game, a test of instinct and wiliness. When she loses the scent, or else the interest, she returns to my side and watches me with shining brown eyes.

I cradle the log in my arms. It's an old stringybark, dark and rough and smelling of earth and damp. The bees have done well to seal it. When I peer into one end, the tiny bluey black bodies, no doubt irritated at having lost their spot in the trees and frantic to protect themselves, squirm crazily over the resin pots, full and ripe and ready to harvest.

A line of green ants march with their saucy orange legs and bold black eyes over the bark. They peer into cracks and jerk upright, their front legs poised to attack, when the log moves in my arms.

Taffy chases a plover over the lawn, the high-pitched chatter of the plucky bird protesting the disruption, and then she meets me behind the shed, where I carefully lay the log on the patch of earth showing through the worn grass.

The shed is small and cramped and smells of engine oil. The spare bricks huddle behind the push-mower, and I jostle with the handle as it whacks me on the shoulder, and shove away the roller in a fit of impatience. The mower clangs against the shovel and sends a domino of gardening tools to the concrete floor. I yank two, then four, bricks from the mess and storm out of the shed, annoyed at my own lack of organization.

Taffy barks at the ruckus as I slam shut the metal door and nurse the bricks back to the bee lot, the soft hum calming my nerves as I set the bricks down. I'll need another pan with water. I've used all the useful receptacles from around the farm: the garbage can lids; the rusty steel drums; even the baking pan that's far too big for the one-person meals I cook nowadays.

Returning to the shed, I extract the spade and sheet of plastic, saved from the purchase of a queen-sized mattress three years ago, when Robert left me. I'd prefer the protective pool to be raised so as not to fill with dirt and leaves, easy to change the water and move the hive, but I haven't the patience to venture to town, nor the money to buy a suitable tub from Hal's Hardware.

It takes me an hour to dig out the hard earth. As I'm fitting the plastic sheet, I hear the hum of a vehicle navigating round the potholes in the driveway. I'm relieved to see it's not the man from the bank, but the familiar Forerunner belonging to my friend and neighbor. The car stops with a jerk, and Carly's flamboyance flounces from the driver's seat. I smile at the tropical moo moo dress over tights and boots and the sun hat decorated with purple sequins.

"What you doing?" she chirps from the gate by the house.

I rest on my haunches and wave. "You're just in time."

She parks her yoga-fit body in front of me, hands on hips, smile shining out from beneath the hat. After a second's assessment, she knows exactly what's going on. She pulls the opposite end of the plastic over the shallow hole and weighs it down with rocks. She eyes the hive. "Storm dropped another log?"

"It'll be fine."

Six buckets of water and a stack of bricks later, the hive hangs in the air like a sculpture, safe from marauding green ants. "It'd better come good."

I stand back to inspect our handy work. "The sugarbag's worth twice as much as the other stuff."

"That bank still on your back?"

"I'm hoping they'll give me more time."

Carly sighs. "Rae, you need to do more than sell honey to keep your farm."

"The flowers are almost ready; Mother's Day coming up."

"Come on, Rae. I know you're not that naïve."

"I just need a few thou' to see me through."

"And then what? Next month will be the same thing. You need to be proactive; get a job, get a man…"

I laugh dryly. "You think a man is going to solve my financial problems?"

She smirks back at me. "He might solve something."

"Meaning?" I cross my arms playfully.

Carly flaps her arms in a surprising outburst. "All you do is mope around here, with your blasted flowers and bees—"

"I have Taffy."

"… and damn dog. When are you going to rejoin society? Your mom—"

"… is gone, Robert's gone. What do you expect me to do?"

"I expect you to *live*. It's not healthy to spend so much time alone. Geoff—"

"Oh, please," I groan.

"He's a great guy. Why not give him a chance?"

I wheel away and head for the house. "I'm sure he wouldn't be interested in a middle-aged divorcée."

Carly trots after me until she is a step ahead. "Are you listening to yourself? You're only thirty-six."

"Middle-aged, like I said."

Carly clicks her tongue and opens the house gate in front of me.

"Anyway," I continue as I head up the porch steps, "At the moment I have more pressing issues than finding a date, in case you hadn't heard."

"Your mom would have wanted—"

"I'm not selling her things, Carly."

"But her collections—"

"No, Carly, and that's final." I want her to leave me alone, to stop hassling me about how I should live my life.

"So what are you going to do, haul all those gems and coins and goodness' knows what around the countryside as a homeless person?"

I hate it when she's so honest and logical and… right. "The honey will see me through this month."

"Even the bees have to give up something they love to survive, Rae."

At the rickety front door, Carly declines the gesture to come inside. "Need to pick James up from basketball. I'll drop in tomorrow, see how the new hive's going." She flutters a wave goodbye and bounces back to the Forerunner.

I watch the car turn in the dirt and drive away before going inside. I wonder where Taffy is. She usually follows me in and sits at the head of the kitchen table to watch me make tea, hopeful of scones or cookies or honey on toast. Today I don't even flick the kettle's switch. I just sit, alone, wondering what to do.

I recognize Taffy's playful bark, and I guess she's chasing the nesting plover, but I can't hear the plover's squawk. When I head back outside, I pause at the house gate and spot Taffy by the shed, shaking herself in an elaborate body wiggle. Why is she wet? I wonder.

And then I see the plastic strewn across the grass and the tumbled log, askew on fallen bricks, one end dipped in a puddle of mud. As I run to the bee lot, I know what I'll find—a deserted log, soggy and lifeless, useless for anything.

I'll harvest what I can, sell it to Mabel's for a good price, but the bees are gone, the hive is dead.

Taffy disappears, knowing she's in trouble, and not yet hungry enough to accompany me back to the house regardless of the punishment.

I head straight to the spare room, my mother's old room, slump on the bed, and cry. For the bees, for me, for how my life went from perfect to pathetic so easily. I can't even protect a log full of stingless bees, and worst of all, I still get their precious honey.

The top drawer of the stained oak wardrobe rumbles open with a gentle tug. It's packed with plastic wallets full of coins. Through the blur of tears, I lift out the first sheet and lay it carefully on the bed. In a silent prayer, I try to explain to Mom: we all need to give up something we love to survive. I hope she'll understand.

Winter Hive
Andrea Dickens

The ball of bees spills across the frames,
seeps deep into the brood box. Their roiling tells me

they're still waiting for warmer days. The snow
piled around the hive has melted, although the earth

still lacks spring's smell. It's like I'm checking,
from the doorway, for your baby breaths, your

whispered sleep. Not wanting to disturb you, but
wanting to be sure of something I can't explain.

The thin crack of propolis popping as I open
the lid. I feel guilty knowing I'm destroying

their warmth, making them work to repair
their home. And at the meager end of the year when

they are beginning to raise new brood. I've no real
reason to open the hive, to worry about them. They weather

the winter together well. But I'm straining across
our distance these days. Yesterday, I saw the first glitter

of wings at the entrance, scouts testing their wings
against the short warmth of afternoon. Watching the bees

climb over one another inside the hive, makes me ache
for those days when I could feel your thrum as I sat in the sun

outside, waiting through summer, my stomach
swollen in time like the squash the bees darted among.

DEFENSIVE
BEHAVIOR

The Bee Bully
The Beekeeper's Scourge
Inga Harris

Burrowing deep into the cell
of your developing babies,
I lay my nest which soon hatches
and feeds from the inside out.
Fresh meat.
As a result, your children may be born
deformed, their wings crippled,
their very presence rendered useless,
soon to be discarded and to die alone
apart from the brood.
My sons sacrifice their lives
burrowed inside yours
but my daughters will thrive
on your worker's blood,
carrying with them the ammunition
of death. They are silent killers.
Allow me to introduce myself:
I am *Varroa destructor*.

Beekeeping with a Honeybee Allergy
Jennifer Ford

When I talk to people about beekeeping, something that always comes up is getting stung and having an allergic reaction. When I first started beekeeping I had what I considered to be "normal" reactions to bee stings—a little redness and swelling, but nothing to really worry about. I would always tell people that while I try to avoid it, getting stung is really no big deal. Well, getting stung became a much bigger deal to me when a few years into my beekeeping adventures I developed a true honeybee venom allergy. Luckily, I have been able to continue my work as a beekeeper.

A few years after my husband and I started beekeeping, we got a call from a friend asking if we could help remove a colony of bees that had taken up residence in an old playhouse. We had never done a bee removal before, but after doing some reading on the subject and getting some advice from an experienced beekeeper, we decided to give it a try. It was an extremely hot summer day and we were moving cautiously, taking our time. By the end, both the humans and the bees were tired, hot, and grouchy. While we were finishing up, I received my first sting of the day—on my right leg. In hindsight, I should have walked away and cleaned out the first sting right away. However, I was focused on getting the job done and continued working. I received two more stings in the same area. I had never had a problem with honeybee stings before, so I was not too worried about it.

We finished the job, packed up, and headed home. As we were driving, I realized I did not feel right. I was breathing just fine but felt nervous and anxious. I then noticed that my lips, tongue, and throat felt slightly swollen. My husband and I talked about it, and he suggested we go to the nearest emergency room. I considered it, but because my breathing was not affected, decided I just wanted to go home. Luckily, the reaction did not get any worse and I felt fine when I got up the next morning. I did some reading, and discovered that true allergic reactions to honeybee venom are fairly rare. I chalked the reaction up to receiving multiple stings and decided to just be more careful.

A few days later we went back to the playhouse to collect any remaining bees. I was stung again and, again, had an unsettling reaction—anxiety, swollen lips, tongue and throat, but no problems with my breathing.

By now I was very concerned about these reactions, and decided to visit an allergist. During my initial visit, the allergist said she sees many bee-keepers about honey bee venom reactions every year. After listening to my story and questioning me about the reaction, she scheduled me to come back for allergy tests. The testing took about half a day and consisted of skin tests of different types of stinging insect venom. For the first few rounds, I exhibited no reaction to any of the venoms. However, on the last round of tests with the highest doses of venom being used, I did have a reaction to one of the venoms being tested. It turned out that I had developed an allergy to honeybee venom. Luckily, I had experienced a relatively mild reaction.

At this point, I was prepared to hear that I would have to give up bee-keeping. I started strategizing ways that my husband could take charge of the actual beekeeping while I helped with bottling honey, candle making, etc. The thought of having to give up beekeeping upset me greatly.

The follow up meeting with the allergist alleviated some of my worries. To my surprise, she did not even suggest that I would have give up bee-keeping if I didn't want to. The bad news was that with every subsequent sting, there was a good chance that my reaction would worsen. The good news was that if I didn't want to give up beekeeping, there were things I could do to make it safer for me.

One was to use more protective gear to avoid stings. For me this meant using coverall pants as well as a hooded jacket, and using gloves when in the past I had preferred to work with bare hands. I was also told that I should carry an Epi-pen with me in case of a more serious reaction.

In addition to the above recommendations, the allergist also suggested that I consider immunotherapy. It would involve getting a shot of a minute dose of honeybee venom, gradually increasing the dosage as I built up a tolerance to the venom. This could help prevent future reactions to honeybee stings, and also prevent my allergy from worsening. The idea of preventing future reactions so I could safely continue beekeeping made me decide to give it a try.

I began going in once a week for three shots of a very minute dose of honey bee venom. I was monitored in the office for 30 minutes after each shot for any adverse reaction. While this was a large time commitment, it was definitely worth it to me to be able to continue keeping bees. Eventually, I worked my way down to two, then one shot with each weekly visit. Then, I was able to increase the time between visits to every two weeks, then three weeks, and so on. While I was instructed to avoid being stung, I did eventually get stung again and had no adverse reaction at all—what a relief! The treatment is not very painful—no worse than a bee sting. I always felt very safe as I was being monitored and never had an adverse reaction to the shots. According to the literature I was given, the shots are 97% effective and most people can discontinue the shots after 3-5 years.

After a while, I had worked my way up to one shot every six weeks, and had pretty much stopped worrying about the allergic reactions. Then one evening, my husband and I went to pick up a nuc that we had purchased from a friend. While we were transferring the frames from one nuc box to another, I was stung and had a similar reaction to my original allergic reaction. The next day I scheduled a visit with my allergist. After meeting with me, my allergist decided to drop me back to one shot every four weeks instead of every six weeks. It also means that I will probably need to continue these shots as long as I continue to keep bees.

Again, this does involve a time commitment, but it is worth it to me to be able to continue beekeeping and feel safe in my beeyard. The treatments may not be for everyone, but for me it has worked wonders. If you have had a bad experience with bee stings, I highly recommend seeing a doctor to find out what might work for you.

Strange Bees
Karen J. Bryant

Mr. Joe had decided to expand his beehives this year. Fred from the Lone Star community told him about a place where he could get a good deal on those bees. Ol' Joe was excited, of course. He was going to add about 100 more hives to the 300 that he already had.

Saturday morning Joe and I headed to Monroeville, which was about 30 minutes from where we lived, to pick up these bees that Fred told about, which sounded like a great deal. Which he told us that we had to make sure we were there early because at this price they'd go pretty quickly. We got there and saw that we were not even close to being the first arrivals. Many people were there, standing in line. We had to wait a few hours before we could even get a good look at the bees.

We ended up taking 250 beehives home, which was 150 more than we anticipated. The price was just too good to pass up.

As soon as we made it to the house, we started setting up the hives. During the setup, I noticed that there were two sets of bees that had no stripes. They were either solid yellow or solid black. After I told Joe, we called Fred over and asked him if he had ever witnessed anything like it before. His exact words were, "Heck, no." I started to worry a little because now I wasn't even sure they were actually honey bees. These bees wouldn't make us any money, and we couldn't afford to lose any money. I told Fred and Joe that maybe that was why they were so cheap. They agreed. Looks like we had been bamboozled. Despite the abnormalities of the bees, Ol' Joe decided to keep them anyway, just to see what would happen.

Word had spread throughout the bee community. People started coming from miles around just to see the strange looking bees. Many of them were making jokes and saying that my hives weren't gonna help me win the "best honey" contest this year. Maybe they were right, but I still had my good beehives left, and that would hopefully be good enough.

After a few weeks, Joe started noticing a difference in the combs he got from the strange-colored bees. The honey that these bees produced was a bright golden honey that almost glittered in the sunlight. He touched his finger to the honey comb and brought it on his tongue. He was in shock at the sweetness of the honey that the bees had made. Never in all his years had he tasted raw honey so sweet. He decided he would put these hives in the back yard where no one could see them and surprise everyone at the contest next weekend.

Even when Fred came by, he didn't tell him about this new honey. He wanted to make sure everyone was surprised. They even produced twice as much honey as normal bees. We had already gathered 400 jars of honey and 250 honey combs to take to the contest. People were going to be amazed at the taste of this honey.

When it was time to go to the contest, I had gathered 800 jars and 550 combs. More than anyone has ever taken before. These bees were super bees, and to think, everyone had made fun of them. What will they have to say now?

I waited until everyone else had placed their honey on the table before I brought out one jar and one honey comb and placed it for the judges to taste. I could see people starting to look at mine and I could tell that they were whispering something to each other. I was starting to feel unsure of myself, but then Fred came up to tell us and what they were saying. I couldn't believe it! They thought we had added yellow coloring to the honey. They even said that we would be disqualified if they found out that it was true. Fred asked if that is what we did. I was appalled. "Of course not!" I told him. "The bees are special and they produced an amazing honey. I cant wait for everyone to taste it."

The judges made it to our honey. Joe and I looked at each other as shocked expressions came across their faces. They came to a unanimous decision and decided that without trying anymore honey, they had found their winners. That did not make the other contestants happy. They were all murmuring angrily until one of the contestants tried the honey and told everyone else to come try it. The angry murmurs turned into praise. Everyone had questions about how our honey could taste so wonderful, and how it was so different. Joe told them to wait while he went to get something. He brought back a few of the bees in a jar. Some were pure black with their eyes just as black, and the gold ones were golden with beautiful golden eyes.

Everyone looked at the bees and wondered where they came from. They also wondered if anybody else that day had gotten anymore like them. We didn't know the answers to those questions. All we knew was that they are special and that nobody else we knew had any like them.

Everyone was asking Joe where he got the bees. He told them and they all wanted him to sell them some. He didn't want to sell any of his bees, even though they were offering big money for them. No one had ever seen anything like them before. Fifteen minutes after Joe put out all the jars and honey combs he had brought, they were gone. He had won $500 for the contest and made several hundred more with his honey.

Once we got back home, we realized what a gold mine we had here with these unique bees. Joe could never sell any of these bees. They were his, and their honey was going to make him money.

Later that night, Joe was suddenly awakened by his dogs barking outside. He woke me up and we quickly ran outside and shined the flashlight into the night. We could barely make out someone running across the yard. Joe immediately thought about his bees. He just knew someone was trying to steal them. He couldn't help but think about how crazy it was to have to worry about someone stealing his bees. He had never had this problem before.

I took off across the yard to see if I could figure out who could have been the person trying to steal the bees, but they were long gone. After checking the hives, I saw that none of them were missing. Our good, ol' trusty dog, Charley, kept the bees from getting stolen.

Joe said he would have to keep them nearer to the house so we could keep a closer watch on them and tie Charley close to them so he could warn us when anyone was near. I just can't believe someone could stoop that low and try to steal our bees.

The next day Fred came by and I told him about what happened. He wasn't surprised. He told us that those bees and that honey had become the talk of the community. I had no idea what we were going to do about this problem. We came to the conclusion that we would have to put an alarm up. Apparently we had something that too many people wanted, and we had to take measures to keep those bees safe.

Things have gotten crazy. At least one of us has to be at the house at all times. Neither of us has much of a life, especially Joe. We have constant calls from people calling and offering insane amounts of money to get some of the bees. It has come to the point where we may have to move, and we never would have dreamed that it would have come to this. We may have to buy a cage and put them all in. I checked on the cage for them. They were very expensive.

One night, Joe and our family had to leave the house. We didn't have a choice. An aunt had passed away, and we had to go to her funeral. Joe left our two Rottweilers to keep guard while we were away. To our surprise, when we had returned home, we found the Rottweilers were gone and all of the hives had been taken. This could only mean one thing: More than one person had to have done this. It would have had to be about three or four people. Joe called the police and reported the crime.

The cops told them that they can't hide long and that they would find them. It has to be someone who was at the contest or who was in it running against you. We will compile a list and start the search. Joe let them know that if they find the culprits, they will brag about their steal. So word will get back to us.

It didn't even take a full day to catch the thieves. I remember seeing the guys at the contest, but I had never seen them before that day. The Sheriff told them that he can honestly say that he had never known anyone to get arrested for stealing bees and beehives before. It was a first for them all.

When the time came for court, the judge wasn't exactly sure how to sentence the bee thieves. I explained to him about how these bees are priceless and important, how good their honey was. Joe and I both explained that they knew how special these bees were and that is why they took them. The judge explained that he could not come to a verdict and pass sentencing until he had a taste of the honey himself. He tasted it and, at that moment, the thieves regretted that they ever stole the bees. They were sentenced to eighteen months in jail and two years probation. Joe and I felt like it was a fair sentence. After much consideration, we decided to sell the bees. Neither of us thought we could live a life that revolved around the bees, and we knew that someone would buy them. They wouldn't be able to resist.

Apiary in May
Payton Sullivan

The fresh spring breeze whipped through Sasha's hair, twisting and knotting the blond strands. The air was warm and the grass transitioning from winter-brown into vibrant green. All along Carniolan Street, children dragged their feet, eager to prolong the morning walk to school. On any other day, Sasha would have set out for school ten minutes early, giving herself time to take the longer route, passing by the Longriver Apiary where she would have paused to enjoy the aroma of wildflowers and to hear the humming of the bees. On any other day, Sasha wouldn't have minded the wind tugging at her hair or the rich scent of thawing dirt. But this morning, Sasha hurried down Carniolan Street, head down and arms crossed, dreading the day before her.

Those who knew Sasha Trench well—her parents, her teachers, what true friends a ten-year-old can be said to have—would describe her as an outgoing child. They knew her to be the brightest fourth grader at Longriver Elementary School and it was tacitly agreed that she was, though still young, destined to go far. Sasha herself, as ten-year-olds most often do, was indifferent to what she was "destined" to achieve. All she knew was that, recently, her charmed and contented life had been abruptly disturbed by that eternal nemesis of all youngsters. Her tormentor, Ella Bankston, was the average fourth grade bully, if there is such a thing. Ella was vindictive, clever, ruthless, and delighted in watching her self-conscious and vulnerable classmates shrink in fear. Her most recent victim was Sasha, whom she mocked to no end, insulting her voice, her clothes, her family, and her friends. After weeks of relentless teasing, the constant barrage of insults left Sasha alone, her friends having deserted her for fear that they would be Ella's next victims. Children are not given enough credit for the damage they can do to one another, targeting each other's flaws and uncertainty. Within weeks, the animated and brainy Sasha withdrew into herself, refusing to seek help, suffering silently as Ella grew more and more ferocious.

This late-May school day was especially rough for Sasha. Ella mocked her skirt in math, ensured that she was last to be chosen during gym, and tripped her in the cafeteria—the most malicious and cruel trick. It was no surprise that tears spilled from Sasha's deep brown eyes as she fled school that afternoon, her backpack bouncing and favorite skirt drenched in pizza grease. Sasha was going to crack. She couldn't take the pain and humiliation anymore and she couldn't return home with a smile plastered on her face, insisting that her day had been amazing, that her grades were still high, and that the reason her friends didn't come over anymore was simply because they were busy. For the ten-year-old girl, it felt as if the world had turned on her, as if the universe itself had rejected her.

Maybe it was the terror that she had glimpsed on her prey's face that day, or perhaps the tears she had noted during their final class, but something that day possessed Ella to, instead of setting off down Chester Street, follow the quickly retreating Sasha down Carniolan. Like a snake relies on the sun to keep its blood hot and flowing, Ella drew her strength from the weakness of others. No matter her reason, Ella quickened her pace, a callous smile staining her virtuous features.

"Hey! Hey, Sasha! C'mon look at me," taunted the pursuer.

The panic that swelled in Sasha's chest upon spotting her pursuer was unprecedented. Her mantra of *sticks and stones…sticks and stones* failed to pacify her bounding heart, and she sped up to a jog. Ella was closing in: thirty feet…now twenty…fifteen. Sasha chose the only action she deemed logical: she dove suddenly through a gap in the chain-link fence bordering the Longriver Apiary.

Ella dropped her backpack and lunchbox on the sidewalk, and squeezed through the fence as well. Sasha was a good fifty feet ahead of her, darting among the nondescript grey boxes. She wove between them, intent on reaching the far gate, which would allow her to exit across the street from her house. Without the paraphernalia of school supplies weighing her down, Ella was rapidly closing the gap between them. She could hear Sasha's terrified gasps and howled in delight when watching her stumble. Within seconds, she caught Sasha's arm in a vice-like grip, and spun her around so they were face to face.

"Let go of me! Let me go, Ella. Ella, please!" cried Sasha, tears spilling down her cheeks.

"What's your problem, anyway?" spat Ella. "It's like you can't even take a joke. Like you think—"

But Sasha wasn't going to take it. Not in *her* garden. Not here, in *her* sanctuary, the sounds and smells and calming atmosphere of which Ella had stolen, along with everything else. Sasha shoved Ella with all her might, toppling her over, into a box. She then took off running, distantly aware of stinging behind her ear on her arms and legs. Not until she had reached the gate did she turn to look back.

Ella and the hive had disappeared in a cloud of honeybees and dust. She was screaming incoherently, a mere formation of swarming insects. Sasha knew what was right. She could picture herself sprinting out of the garden, though her own front door, into the kitchen. She could picture herself distraught, insisting that her mother telephone for an ambulance! A girl was dying in the bee garden! She's allergic to bees, you know. She's got emergency epinephrine in her backpack everywhere she goes. Hurry, she'll be dead in seconds!

As Sasha surveyed the struggling Ella, every sour tear, every lost friend, and every suppressed retort shattered within her. And so, she turned her back and darted out of the apiary, where she collapsed on the sidewalk, hands clasped over her mouth in fear that her overwhelming knowledge of right and wrong would betray her. Gradually, the cries on the other side of the fence faded. Painfully, Sasha picked herself up.

Drivers passing by gaped. They slowed their cars and craned their necks in order to study the ruin of a child swaying on the curbside. Her hair was tangled and her skirt stained. Welts were swelling on her arms and face. Her tights were ripped and her knees bleeding. Her countenance was tear-streaked, devastated. Sasha was screaming, screaming, screaming inside. The distant aroma of wildflowers wasn't sweet anymore. The muted hum of honeybees didn't seemed more like a roar than a hum. The garden was ruined.

Diablo Canyon | Alice Many Ears of Corn
Gary Ives

This story was told to my mother by her mother. A long time ago, the place now called Diablo Canyon had belonged to our people, the Chiracahua. The Apache name for the place *Ch'inodaah Hela* translates roughly as "Don't come out." Although there is abundant water in this canyon, no one lived there as it was considered a place of witches. Anyone entering that place might come out crazy or stupid. Hunting parties skirted the north side of the canyon but dared not set foot on the south side, and never camped overnight in the canyon. Kukuk, a great hunter, a warrior and a shaman killed many Mexicans and Americans there. Several times, gold seeking Mexicans were led or pursued into this hidden canyon by him. Our warriors had only to wait until those Mexicans or Americans stumbled out like drunken fools, then easily broke open their heads with rocks.

Those first Mexicans came long before the time of Kukuk. Strangers from the south, they were killed, but more came. They too were killed, but still more came. In those days, our people had neither horses nor guns so the Mexicans with their horses and guns were able to kill many of our people, steal our women and our food and burn our villages. However once we too had horses and guns, we did the same to them. Eventually the Mexicans came to respect our lands and stayed over there in their places. Occasionally, raiders violated our lands as we did theirs but not many people were killed or taken and few villages were burned. Those Mexicans would come looking for gold or silver, while our warriors went to steal their horses, corn, and women. Many of our mothers were taken from the Mexicans. My grandmother, she was called Ekta, was taken from the Mexicans when she had but twelve summers. This is how many of our people came to speak the Mexican tongue.

The Americans came and beat down those Mexicans, pushing them further south, then turned on our people. From the Jicarilla, from the

Mescalero, from the Comanche and the Navajo we knew of the greed of advancing Americans with their cattle and their blue coats arriving like grasshoppers coming out of the ground, their numbers impossible to count. Eventually they pushed our people out of the valleys and into the deserts and mountains where life is hard. The blue coats had rifles that fired bullets without needing to be reloaded. Worse were their giant thunder guns, canons, which made huge rocks explode far away. After the Americans killed Mangas Colorados there was much concern that the soldiers might also come to take the mountains from us. Fearing such, a small band from the Juniper Clan, my grandmother and grand-father among them, risked moving into Diablo Canyon, that canyon of witches. Numerous springs dotted the south face of the canyon high and low. There on the south wall, vegetation and the animals were peculiar. Jaguar and coati mundi were often seen and a kind of tan colored bear of which there are no more, and there were large, slow, spotted lizards. Among the wild azaleas and rhododendrons which grew in profusion were great numbers of a kind of bee which built hives in the open, hang-ing low from branches. So many of these bees' hives existed that steal-ing the honey was easily done with smoke. However my people quick-ly learned that whoever ate this honey became crazy. A person's mind could be disturbed for one day or even permanently depending on how much honey he ate; the honey held powerful hallucinogens. Once our people learned that the honey was the source of craziness, life in Diablo Canyon was good. Occasionally Americans looking for gold wandered into the canyon. Our people killed them and ate their mules.

Even more Americans came after their war in the East. These Americans were quick to grab land, especially land with water. A greedy speculator who had spotted the greenery of the canyon filed a claim with the land office. He then sold our canyon to a group of Christians who planned to settle there and to convert our people to their way. There were about seventy of them who moved into the canyon with many wagons loaded with food, tools, and lumber to build their houses and church. Quickly, they found the abundance of bees and set about robbing the hives. Our people watched as these Americans quickly became crazy drunk on the mad honey. Some of our women made wicked masks which they put onto sticks and poles and they howled like wolves in the night to further frighten these settlers who were too afraid or too crazy to sleep. Through the night they drank coffee and tea into which they added more and more honey. By morning those not dead were completely crazy, stumbling around their camp of wagons like blind people. Some of the armed men who had survived the night shot each other with pistols. The two women and several children who found their way out of the canyon never recovered. The Americans named our place Diablo Canyon and left our people in peace for many years until the One Horse Mining and Copper Company came with their miners.

This is the story my grandmother told to my mother.

THE BUZZ

Tale of the Dragon Bee

Randy Ames

Old Granny Kala leaned forward in her rocker, the one her equally old man had made for her.

"Gather round, children, and I will tell you The Tale of the Dragon Bee."

Unbeknownst to Granny Kala, her man walked in quietly from the kitchen to stand behind her as she began her tale. His hair was just as white as hers and almost as long. He put his finger to his lips to hush the children to his presence.

"Our town thrives on the wonderful honey made by our extraordinary honey bees. Our bees are not magical, but they are larger than those of other towns in the valley." Granny Kala said with a flourish of her right arm fanning out to indicate the other four towns in the Valley of Sprinkling Water. "The Queens are the size of a thumb."

Granny Kala's man held out his two thumbs and wiggled them up and down. Three of the children giggled, but he put his finger to his lips to quiet them to keep Granny Kala from noticing him.

"Now we tend to the clover growing in the fields so that our wonderful bees can collect the best pollen and nectar, and we maintain the hives so that they can make the best honey in the valley, but the Queen Bees govern their own drones and workers themselves." Granny Kala said with a smile. She had told this story a hundred times in the first half of the year alone, but the children never got tired of it, or the beekeeping lesson they learn from it.

Granny Kala's man tucked his hands under his armpit and flapped his elbows. Five children giggled at his antics.

"It was mid-summer when the Dragon Bee first came into our valley and attacked the hives." Granny Kala said with a terrible treble in her voice.

Two of the children covered their eyes.

"The Dragon Bee flew into our valley from the south end. We heard reports from other villages as to the terror the Dragon Bee was creating." Granny Kala paused. "The Dragon Bee was unstoppable. Every hive it attacked was left deserted and devastated. Finally it flew into our fields."

The children lifted up their faces in awe and suspense.

"The Dragon Bee was bigger by far than even our Queen Bees…" Granny Kala paused for dramatic effect.

"It was as big as a house!" Interrupted Granny Kala's man with his arms flung out wide.

A collective gasp erupted from the children. Granny Kala turned her white haired head to frown at her man.

"No it wasn't!" Granny Kala rebuked him. "It was the size of two thumbs side by side."

Turning back to the faces of her rapt audience, Granny Kala continued. "But it did breathe fire; a red hot flame as long as-."

"As long as a tree!" Granny Kala's man exclaimed.

"No it wasn't!" Granny Kala said in frustration. "It was as long as your arm."

Granny Kala swung a hand at her man, but he danced out of the way. Half of the children laughed out loud.

"The Dragon Bee was so large and powerful that the buzz of its enormous wings could be heard a mile away." Granny Kala frowned to show the children how sad the next part was.

"We have ten hives, with ten Queen Bees, and thousands of drones and workers, but they were no match for the ferociousness of the Dragon Bee. He came in from the south woods, flaming the tender clover blossoms to ash. Then he reached the first hive. The Dragon Bee attacked without provocation. He dove at the hive and breathed his deadly flame on it. The wood was singed and scorched, but it held.

Granny Kala relished the eager faces of the children in rapt attention to her every word.

"The Queen Bee gave her orders; the workers flew out to defend and attack. The Dragon Bee would be overwhelmed, as it was when wasps attacked. Workers and drones would die, but the hive would be safe and the Queen had enough stored sperm to produce enough females to re-grow the colony and collect more pollen and nectar.

"The Dragon Bee could not be overwhelmed. Workers fell in droves. Thousands of charred bodies littered the space about the hive. Eventually there were no females left to sting. The drones, brave as they were could not hope to defeat the Dragon Bee with no stingers, but even they flew out at their Queen Bee's command.

Granny Kala's man sat down beside her in his own rocking chair. She continued the tale. "Hive after hive fell to the onslaught of the dangerous and dastardly Dragon Bee. Before the villagers had realized it there were nine smoking ruins of wood and wax. There was only one hive left. The Dragon Bee stormed straight to it."

The children gasped in anticipation.

"As the Dragon Bee approached to destroy its last target in the valley, no swarm came from it to fight him. No workers with their tiny stingers, no drones with their thicker bodies and useless attack, only the last of the Queen Bees."

"The last of the Queen Bees was brave and smart and had seen what had befallen the other hives when their Queen Bees acted naturally to defeat the terror that assaulted them. This Queen Bee thought differently and came out alone to face the terrorizing insect. The Dragon Bee halted in mid-flight. Hovering, the Dragon Bee was dumbfounded. Then he dove in for the attack. The tremendous buzzing of his wings and the glistening tip of his stinger should have drove the last remaining Queen Bee to flee, but she stayed the course."

"For five minutes the Dragon Bee dove and strafed with his long flame, and tried his very best to kill the brave little Queen Bee. The Queen Bee was smart and instead of attacking the Dragon Bee she avoided the sting and fire."

Granny Kala stopped for a long breath. The children leaned forwards, two holding themselves up with their chubby arms.

"And do you know what happened then?" Granny Kala asked: the children nodded, but were silent. Even Granny Kala's man stopped talking.

"She kissed him. The Queen Bee dove at the Dragon Bee, flung her legs about him, pinning his wings and while she held them both aloft, she kissed him."

Three of the boys made faces, while all of the girls, smiled and laughed. Granny Kala's man finished the tale.

"When the last and most resourceful of the Queen Bees kissed the Dragon Bee, his fiery heart melted, and he kissed her back. His flame retarded, and from then on he only blew smoke from his nostrils. Lying on the ground, with the Queen Bee entwined about him, he became her King and from then on he did everything she wanted him to do.

Granny Kala turned and kissed her man, and, of course, the children went, "Ewwww."

From a Bear's Perspective

Josh Wachtenheim

Oh, honey bee! Oh, honey bee.
Where would your honey be?
I do not have a money tree.
So can I have some honey, please?
I admire your work
in your hive where you lurk!
So don't go berserk
because it comes with some perks!
The delicious, sugary,
mouthwatering honey.
That is the prize
that comes from the hive.
So I hope you won't care,
I'm sure you have to spare,
if I took some honey from there
back to my lair.
And I'll bid you adieu
plus a warm "thank you!"
for the honey that you—
you—
Ah-choo!
It seems, sadly,
how unfortunately…
Oh, how could this be?
I am allergic to honey!

Regibald Sees the Eggs

William Blomstedt

"Well, I'll be pickled," Regibald Winklesteep said to himself, pinching a stinger from his hand. He hadn't seen the bees this unhappy for a long time. And he'd been a beekeeper a long time. "What's wrong with ya?" he grumbled. Normally he had a nice bunch of bees. When he'd sold off the big chunk of them at the end of his commercial days—15 years ago now—he'd only kept the gentlest, most productive colonies. But at certain times, this being one of them, the bees still showed that streak of nastiness he thought he'd gotten rid of decades ago. This hive, his third of the morning, was the real bother. Merely cracking the lid sent him back to the truck for a veil.

"Bees, bees, oh boy, oh bees," Regibald said. Mad as this hive was, he was determined to lay eyes on their queen. He gave them another puff of smoke and then jammed his hive tool under the super. With the good horsemint flow this spring the super was already heavy enough to make his lower back sing, but he'd lifted heavier in his day. Underneath the super, the nurse bees poked their heads into the new light. He sent a puff of smoke across the top-bars and they responded with a roaring buzz.

Regibald pried out one side comb and lifted it into the air. They had a good, thick strip of pollen but the brood looked a bit spotty. He brought the comb to his face and peered closely in between the capped brood. Even when it was sunny he had a bit of trouble seeing eggs, but today was cloudy, and his poor old eyes could barely see past the veil, much less the small piece of rice at the bottom of each cell. He had a pair of glasses somewhere.

Meanwhile, the bees continued stinging. Setting the comb against the box, Regibald walked back to the truck and found an old pair of gloves smashed behind the seat. On the return, while trying to slide his hands in the stiff leather, a broken piece of wood reached up and caught his foot. He tumbled forward, his arms striking the hive, but instead of rolling off he had the most peculiar feeling: like he was laying on a writhing bed of hard fuzz, and somehow, slowly, moving downwards. The light was dim, but after a few moments he toppled head over heel to the bottom of a small, warm cave.

It seemed a bit strange, to have fallen into a cave in his bee yard he had never seen before, but Regibald had seen lots of strange things in life, so he stood himself up with the normal set of cricks and creeks and brushed off the dust. Only when he took a few steps towards the light did his eyes focus on the head of a giant monster only a few feet away. "Egads!" Regibald shouted, but then recognized it was no monster, but a bee; antenna twitching, mandibles opening and closing, and shaking its head at him in a not-altogether-friendly manner.

"Hey!" Regibald shouted. "What are you so big for?" The bee didn't answer but Regibald realized if it could say anything, it'd probably ask, "Why are you so small?" For that's what he was; tiny enough to stand up in a cell. Somehow he'd fallen into the very beehive he had been working in just minutes ago. This bee across from him was still a pale white-yellow fuzzy baby. It had just poked its head through the wax capping and was probably as confused as he was.

"Don't worry, young miss," Regibald said to the bee. "You'll be with your sisters soon enough. But me? I gotta get out of here. Supper's in an hour." A group of workers walked over the opening of his cell, temporarily blocking the light. One even stuck her head inside and took a look around, but didn't seem to care about him. When the entrance was clear, Regibald climbed to the edge and took off his veil to get a better look. His cell was half-way up a comb which was as tall as a city building. Above him lay capped honey, around him was more brood. Working bees scurried about the wax formations. A big oaf of a drone lumbered nearby, buzzing.

"I'll be plumbed," he said to the baby bee. "You know, a few years ago I could've hopped down there in a jiffy," he said, pointing at the floorboard far below, "But not today. Not with these knees." He looked for another group of bees to ride downwards, but they had likely gone in search of the hive-disrupter.

"Hmm," Regibald said and turned around to see what tools he had on hand. Looking behind him he saw a small pile of white logs in the bottom of the cell. Eggs, he thought, but why so many…

"Ah-ha!" he shouted. "I see the problem." He brought an egg to show the bee-head across from him. "Sorry to say, but you have a laying worker among your sisters." He dropped the egg and watched it tumble to the floorboards. "We'll fix that up later and you'll be as right a rain. But first, I gotta get out of here"

Then Regibald saw a drone the size of a cow trudging towards his cell. Just as it was about to pass, Regibald jumped onto the drone's abdomen, grabbed a handful of fuzz and hung on tightly. The drone gave a mighty buzz, but didn't stop waddling along the comb. Regibald couldn't see much with his face pressed against that hairy cuticle, but he felt the bee walk up, then down, then sideways. It wasn't quite as vigorous as riding a bull, which he once tried back in his foolish days, but it did have a similar sort of excitement when the bee had him hanging completely upside-down. As soon as he felt upright, Regibald eased his butt to the floor and let go. The drone gave a buzz of relief and waddled away, leaving Regibald alone on the bottom board.

"Well, I'll be peared," Regibald said as he swept himself off once again. "It's sure been an odd day" The bottom of the hive was dark, but beyond the clusters of bees he could see daylight through the long, narrow entrance. A more timid man might have crumpled at the thought of the bee's mandibles, now large enough to snap his leg in two, or a stinger the size of a broadsword able to pump a gallon of venom, but Regibald spat on the ground and said "Those bees never pay any attention to the damn hive beetles walking around, so why should they care about me?" and began walking towards the exit.

All the bees were quite busy with their activities. A few seemed curious at his appearance and whenever they touched him with their antennas, Regibald just grabbed one for a firm handshake. Only when he neared the entrance did one bee, perhaps one of the guards, begin to give him a bad eye. When it came a little too close, Regibald yelled "Git!" and gave it a kick. Instead of driving the bee away it seemed to stir up an alarm pheromone, and a few more bees began to give him a bad eye.

"Woah, now," Regibald said. "Remember who left all that nice honey last autumn to get you through the winter?" He edged away from the group. "And who treated you for that case of foulbrood this spring? And gave you that extra frame of eggs?" The bees weren't too keen on his words and right as they looked ready to pounce a big drone buzzed between them, begging for food. With the bees distracted, Regibald bee-lined towards the entrance. With his first steps into the fresh air, the bright light struck him so that his foot slipped and he tumbled off the landing board and went falling, falling, falling down into the grass.

When he came to, Regibald sat up and looked around. Behind him the hive was still open, now its normal size. What just happened? He thought, rubbing his head, which did ache so. He picked himself up from the grass and, through the cricks and creeks, stood up straight.

"I think I've had enough beekeeping for today," he said. "I'll put a rock on it and give the hive a good look-through next time. If I saw a laying worker now, well, I'd probably just head straight to the nut house and check myself in." As he was putting the hive back together, a bee stung him on the cheek. He swiped it off and stood still for a moment. Then he patted his head and looked around the hive. Where was his veil, anyway?

Waggle Dance

Andrea Dickens

My first hives—bees dart in and out. Monday nights: scrape
of Extension Center folding chairs. We're learning how
to keep alive bees, bees who probably could live better without us.
How many honey supers to add, how many we must leave
for winter. After break, we talk purpose: everything to support
the queen. Gather pollen and nectar, warn of danger, nurse
her young. I think of the babysitter, the pile of work I left
at 5, the mortgage, the check engine light, the overdue
library books. I yearn for such simplicity as bees, one task
at a time. The speaker launches into the waggle dance,
their figure-eight that tells direction, distance, a tease with
the promise of nectar. So different from my scrambled to-do lists
as I lurch from day to day. I wonder, watching the video
of a worker dancing: is it okay to be jealous of the bees?

The Bee Thief

MFC Feeley

Harry was back! Fruitus charged up our hill, tail wagging, grass and cosmos flattening in his wake. Dew soaked the hems of my overalls and cut my ankles as I scampered after him.

But the guy standing over our neglected bee box wasn't Harry. He was shorter. Stockier. And there was no way Harry could afford that crazy opalescent bee suit.

This was the Bee-Thief.

I'd thought, as had the police when Janet reported her hive stolen, that the Bee-Thief was only a rumor, an apiary legend striking randomly across the planet, giving shoddy beekeepers an excuse for colony collapse. "No Varroa Mites on my watch; it was the Bee-Thief!" But here he stood. I owed Janet an apology.

I looked down at the road and didn't see any tire tracks in the soft earth, nor was a truck or car parked anywhere. The Bee-Thief hadn't been trampling through the woods. His suit looked immaculate. He seemed to glow in the morning light.

I'd been alone a long time.

Fruitus lay down exactly fifteen feet from the hive and growled. "He always does that." I lied. I wasn't afraid. Yeah, thieving is wrong, but there's plenty right about a guy who can't resist bees. Janet says I'm too trusting. Janet also says I should meet new men, and here's a guy who shares my passion for mellifera! Plus, nothing's happened in my life since Harry left. This was something.

"Need some help?" I asked, stepping closer.

The Bee-Thief threw himself over the hive like an accountant sheltering his computer in an earthquake. Even under the mask, I saw his eyes swell with fear. I switched to the voice I use at the dog shelter when a hound pins his ears. "Shhhh. Shhhhhh." The Bee-Thief relaxed a little, but he didn't let go of my bee box. I snapped my fingers to bring Fruitus to my side. Fruitus whimpered and stayed where he was. Once stung, twice shy. Coward.

Changing color in the light, the Bee-Thief's suit gave the illusion of transparency. Harry would be so jealous. "You don't have to steal," I said, thinking of how expensive that fabric must be. "I could help you. You could come to our meetings." The Bardonia Beekeepers meet once a month. Janet would love this guy (I mean, after he replaced her bees). By now I was a few feet away from the Bee-Thief. I felt the magnetism that passes between the unmated. The morning mist condensed into a light drizzle. I wondered what he did with his stolen bee boxes.

"If you can afford that—" I started, jabbing a finger at his gleaming bee veil. The Bee-Thief hefted up my box and took a step backward. I hadn't meant to accuse. Mirroring his body language, I stepped back too.

"Habla Espanol?" I tried. "Po-Russki? Parley-vous Frances?" What was the harm? I had my dog. The Bee-Thief's hands were busy with fifty pounds of hive. He couldn't hit or grab me. He couldn't draw a gun. He might have dropped the box and sent bees everywhere, but I didn't think of that. My instincts said he was harmless. Maybe a little mentally deficient, but good. He loved bees. "Spekensi Deutch?" I smiled my warmest smile. It began to rain.

The Bee-Thief looked skyward and started hopping from one foot to the other.

"You don't want to do that!" I warned. Vibrations upset bees. I gestured toward the house, "It'll stop in a minute. You can come inside." He kept jumping. "Slow down." I imitated his dance at the pace of a sway, but the Bee-Thief gritted his teeth and stamped harder, shaking the hive, rattling the trays inside, bruising workers, smashing larvae. Fruitus barked. The Bee-Thief jerked around so violently that his veil lifted and caught on his beard.

"Stop!" I shouted. It was too late. Honey bees are non-violent, but even Carniolan will defend their hive. Bees blazed out of the box opening and attacked the gap between the Bee-Thief's veil and his collar.

The Bee-Thief wailed and dropped my hive. It fell on its side, but seemed intact. Something snaked through the sky. I thought it was lightning. Thunder cracked and rain poured which quelled the bees or I don't think the Bee-Thief would have survived. He writhed in anaphylactic seizure. Once muddy, his gleaming suit stuck to and outlined the Bee-Thief's muscles. He was slowing down. Stiffening. I didn't know how much time we had. I'm not allergic to bee stings, but I always bring an Epi-pen with me when I'm working on the hive, just in case. Now it was in my kitchen.

I rolled the Bee-Thief down my hill. Fruitus chased him like a ball and licked his neck. Good. Dog spit is anti–bacterial.

The Bee-Thief was semi-conscious as I dragged him to my couch and pulled off his helmet. "You're a Yeti!" I cried when I saw his fur covered face. I wasn't sure whether his neck was swollen or if it was always that thick, but the red welts blistering his downy cheeks scared me. His breathing grew labored. "A Yeti. That makes so much sense." Except the clothes. I didn't think Yetis wore clothes. I lifted my Epi-pen, looking for a good place to stab him then froze. How would Yeti blood react to human medication?

The topic at our last beekeepers' meeting was "Propolis." Bees repair their hives with it. It's a natural poultice. Janet brought samples for everybody and I still had mine in my coat pocket. I parted the Yeti's beard and picked the dead bees from his long golden hair. I made sure to find one welt for every bee corpse and squeezed stingers out with my fingernails. I rubbed the propolis on his wounds and was amazed how quickly the red calmed. (I am totally going to start harvesting propolis.)

Then I got a jar of the honey Harry and I had extracted the day before he left and spooned a little into the Bee-Thief's mouth. His face was human, but kind of beaky. His nose thrust forward and blended smoothly into his cheeks like a pollen eater or a fleshy duckbill. His general fuzziness looked friendly. And he was handsome. Not because I like hairy weird faces; handsome because he was intrepid. I brushed my fingers through the hair around his face and discovered small clean ears. Really cute.

He started breathing better. I spooned more honey into his mouth. He slurped. Slowly his eyes opened. An impressive golden brown under heavy lids. I forfeited English and just smiled. The Yeti smiled back. The universal language. Fruitus licked his face again and the Yeti laughed.

I couldn't wait to show him to Janet and prove I'm not just sitting around missing Harry. But it wasn't to be. One minute my Yeti was barely awake, patting Fruitus on the head and watching the rain out the window; next moment the sky cleared and he ran out the door.

"Where are you going?" I sprang after him, still holding the honey. Half-way up the hill my Yeti stopped. My heart leapt. He was staying! But he wasn't. My Yeti started his dance again, this time keeping a safe distance from the hive.

Just when I recognized his moves as a waggle dance, like bees do to tell each other where the nectar is, a translucent hose dropped down from the sky. In the post rain brightness, I almost couldn't see it, but as that luminous tube drew nearer I knew I was about to lose my Bee-Thief.

"Wait!" I ran up to him and pressed my honey-jar into his gloved hands. Then I fell back and grabbed fistfuls of grass as a great sucking pulled everything toward the shimmering hose. It vacuumed up my Bee-Thief, then twisted up the hill, swallowed my bee-box, and drifted away.

It's happening all over the planet. It happened to Janet. Our bees are disappearing. Now I know how, but I don't know why. Do the Yeti-men just like honey, or did they maybe, in their quest for technological advance - on their way to intergalactic travel - do something that killed the bees on their planet just like we're doing here?

Maybe we're not killing our bees after all. Maybe it's all Yeti-piracy. Either way, it doesn't matter. Our bees are vanishing. We have to replenish them. So, I'll start a new colony and guard against mites and bacteria, monitor the shade and sunlight. Janet's right. Harry's never coming back, but I hope my Yeti will.

Fred the Bear

Roman Carrier

Deep in the woods, beyond the way,
over the hills in a dank, dark cave,

past the meadow where the sunlight bled
lived a big black bear whose name was Fred

Fred was a bear who was calm but not bold
who feared the coming snow and wintry cold.

The warmth would disappear with the orange sunset,
so Fred had to gather all the food he could get.

A bulk of grub meant for his hibernation
for bears have a long rest during winter's duration.

Fred left his cave in quite the hurry,
determined to find fare before slush and slurry.

After hours of searching, just before hope was lost
What great gift should he come across?

Hung from a tree was a very unique fruit
One that marked the end of this bear's pursuit.

For once he laid his eyes on it he knew
a bountiful feast would come from this vessel of golden goo.

He sauntered over, savoring the harvest of his great snack
When out emerged creatures of yellow and black!

"What business have you, attacking our hive?
Sure as the stripes on my back, you won't leave here alive!"

"I meant no harm," Fred said, "I've got a long sleep.
Could you blame me for seeking some honey to reap?

"Can't you see? I was drawn by the amber glow,
A beacon of hope against the forthcoming snow"

and the bees had a bit of mercy that day,
they left Fred unstung and sent him on his way.

But Fred was still hungry, with no food in sight
Finally he thought give up he just might.

Fred pondered the cost of a bear-sized casket
when past the bush he spotted a picnic basket.

Fred ran as fast as he could, making haste
Sure that his new gift should not go to waste

and just when he grasped it, ready for a taste
A danger even greater than starvation he faced.

No longer worried about the setting sun
for he met a park ranger with a badge and a gun.

The barrel of a gun was the last sight he'd see
Until out of the brush came a swarm of bees!

The ranger would flee while bees pursued,
and in the commotion, he'd forgotten his food.

They chased him out of the woods beyond the way
Over the hills, past a dank, dark cave

After the meadow, where the sunlight bled
and lasted another long winter, did Fred.

King Bee
Lela Marie De La Garza

Charles was in a bad mood that morning, and Mallory made things worse by remarking, "You sure got up on the wrong side of the bed this morning. In fact, you got up on the wrong side of the room." Charles said some unprintable things to her, and she said them right back at him.

Finally he got on his high horse and rode it out the kitchen door and all the way to the apiary. There he sat on a bench and looked at the hives. *Bees are stupid*, he thought. *The worker bees do nothing but work till they die. And the drones—well, they're even worse. All they do is wait around waiting to service the queen. It's like a bum lying on the sidewalk hoping for a half bottle of whisky to drop from the sky. Now the queen is the one that has it made. She produces. Without her there wouldn't be a hive. I wonder why there isn't a king bee?*

Charlie stood up. Half-jokingly, he spoke to the hive: "How would you like a king?" Something sharp stung his ear. He put his hand up and pulled out a tiny gold arrow. There was another sting on his forehead and one on his neck. *They don't want me here*, he thought. "Stop it!" he roared. "I'm your king!"

A large bee flew out of the hive and toward him. Fearing to get stung again, Charles stepped back. But the bee turned into a beautiful woman with amber hair and jet black eyes. She wore a yellow robe, and in her hand was a crown, which she held out to him. "I've thought of abdicating for a long time," she said. "It would be nice to go on a world trip—see beehives in Hong Kong… caves in Egypt where they've found honey jars thousands of years old… France… Germany…

Put this crown on, and you'll turn into a bee. Though laying eggs might be a bit of a problem for you. And I don't exactly how you'll manage to mate with a drone."

144

Charles took another step backward. He hadn't thought that far. "It might not be such a good idea after all," he said.

"In that case, why don't you drink this?" The crown this lovely bee-woman was holding out to him turned into a cup.

Charles looked at it suspiciously. "Will it change me?"

"Only for the better." Charles wasn't sure what that meant, but her beauty was hypnotic. He took the cup and drank.

The honey was sweet with an edge of darkness to it. Everything began to swim, and he felt himself losing consciousness...

***

Charlie woke up in his own bed. For a minute he just lay there, wondering which was the right side. Then he saw that half the room was filled with golden light. He rolled into it and got up.

***

Mallory noticed that Charlie was whistling when he came into the kitchen. "You certainly got up on the right side of the bed this morning. Have some breakfast."

"I will," Charlie said. "But first I'm going out to the apiary and check on the queen bee. She might have decided to abdicate."

Mallory laughed. "That's ridiculous. The queen stays with her hive for life. You know that."

"You're right." Charles sat down, poured a cup of coffee and grabbed a piece of toast.

Mallory looked at him more closely. "There's something shining in your hair. What is it?"

Charles reached to his head and felt something tiny and hard. *No,* he thought. *It was just a dream.* Wasn't it? His fingers fumbled out a minuscule gold arrow. Charles looked at it for a while and then got up. "I think I'll go out at have a quick look at the hives," he said. "Just in case."

ANOTHER PERSPECTIVE

Honeybee's Journey
Abigail Miller

The gentle breeze whispers aloft rolling hills,
Sapphire summer sky,
Tranquil grassy fields
Pedant from a willow, a flourishing honeybee hive
Within the sweet amber cove,
A tiny kingdom of liquid gold
Synchronized humming creates a song
Bees sing as they go along
Intricate art under construction,
Molding wax without disruption
On every cell honeybees dance
Her majesty orders one to advance
She hops out of the nest and flies
Aerodynamically
Translucent wings convey her
Beyond the glistening stream
Bubbling amidst hazel stones,
Beyond the timeless cottage
With ivy draping its eaves,
And finally she arrives
Canvas of color
Wild flowers grow together,
Countless blossoms beyond measure
Painted with shades of violet and red,
She flies down from overhead
Landing softly on a petal,
She gathers some nectar
The sunset is now so magnificent,
Making a blissful gradient
Her velvety coat sheens in the fading light,

It's time to return before the night
She retraces her path,
The whole way back to the hive
The monarch waits patiently to evaluate her prize.

The Tale of the Sting

Tibor Csincsa

At the beginning of the time, the natural world was much more peaceful than today. All creatures lived in harmony and peace without any harassment of predators, parasites and diseases, or even humans. It was a real Paradise.

The bees were hard workers at that time too, long before the human used to call them busy bee. The bees collected nectar and transformed it to honey. They collected pollen to make bee bread. They harvested way more than they needed for daily consumption and rainy days when they couldn't forage. The bees kept a large storage in their nest, because, at that time, man-made hives didn't exist. They needed it because, as opposed to other insects, the bees didn't hibernate or die in winter.

This was where their problem began. Although the animals lived in peace, they weren't free of weaknesses such as envy and a strong desire for sweetness and easy snacks.

The bees had many difficulties to keep the others out of their precious honey because, at that time, they didn't have a sting or any other means to protect themselves or the honey, which was very essential for their winter survival. To stay alive during the coldest months, they needed a lot of energy, and that energy comes from the honey.

The bees decided to visit Mother Nature and complain about their dire situation. "Everybody, from the smallest ants to the mice to the largest bear, is stealing our honey, and we have no weapon or any means to protect our essential winter food," said the bees.

"Hide your honey better," replied Mother Nature.

"We tried to keep our entrance hole to the minimum, but we need some ventilation to ripen our honey, and the ants could enter even the smallest holes. The woodpecker can chisel our nest entrance as large as he wants it."

"Hide your nest under rocks and the woodpecker won't bother you," suggested Mother Nature.

"We've tried that, but the bear can roll over the largest rock to eat our honey," cried the bees.

Finally Mother Nature realized the hard-working bees were right, and they deserved some help. "I can't give you large strong jaws or horns because they will be in your way to nibble the nectar in the small, deep flowers. However, I can give you a needle-like sting which will be on the end of your abdomen. I will make it even a little poisonous so your piercing will be more painful," Mother Nature said and sent the bees away.

The bees were happy. Their uninvited guests were very surprised when they wanted to put their hands on the bees' treasures. Slowly, the animals left the bees alone most of the time since they had learned how painful the sting could be, especially multiple stings on their nose, ears or lips.

After a while the sting-armed bees started to get overconfident or even aggressive, because much larger creatures became afraid of them. They started to use their sting for attacking others too, not only for defending themselves or their honey.

When there was another insect on the flowers that the bees visited, they chased the others away without any delay. Many other insects made a living from flower nectar as well, like flies, butterflies, bumblebees and other bugs. The bees didn't tolerate anybody on their forage ground and wanted to keep every flower to themselves. Not only were the other insects unhappy, many plants were, too, because some flowers can't be pollinated only by bees.

Complaint after complaint, Mother Nature decided to summon the bees and question their aggressive behavior. "I will take back your sting because you abuse its purpose and harass innocent others."

The bees were frightened and immediately started to beg Mother Nature. "Please don't take our sting away! Everybody will retaliate and leave no honey for us for sure. We promise, we are going to use our sting just as a last defense," cried the bees.

"This is what you should have done from the very beginning," replied Mother Nature, but she didn't want to leave the bees without any defense. "I'll make sure you keep your promise," said Mother Nature. "From now on, your stings will have barbs and you can't pull it back after you've pierced somebody. This way, it'll be guaranteed you can only use your sting once in your life, because, when you do, you will die as well."

The bees were breathless from surprise, but Mother Nature left no chance for any other arguments. "I'm finished with you and mind your own business," she said and sent the bees away with a wave of her hand.

The bees went home sad and scared. At their arrival, they realized that Mother Nature wasn't joking. Each of them had a sting with barbs on it. Since then, the bees shaped up and left the other nectar collecting insects alone on the flowers. Even today, the bees use their sting only as a last defense because they know they can sting only once before it's the end of their own life as well.

The Wasp and the Bee

Tibor Csincsa

The wasp was very fond of the sweet, almost addicted to it. In the fall he often pilfered the honeysweet grapes, caramelized plums, aromatic apples and other fruits. However, these traits were available mostly in the autumn and even then for quite a short time. The bulk of the fruits were quickly harvested and the fallen ones on the ground were eaten by the animals rather soon. Summer time he tried in vain, these fruits were absolutely sour and unenjoyable.

Once, in the forest, the wasp was wandering around when suddenly a sweet odor touched his nose. Like a hound on the bloody track, the wasp followed the pleasing smell. Soon he noticed a cavity on a large tree, which was the home of wild honey bees. There was no doubt where the sweet smell was filtering out from. The wasp long observed the unsuspecting bees from his hideout. The bees were coming and going and seemingly didn't pay much attention even to each other. Each bee was busy with her own business.

The wasp couldn't resist the temptation for a long time and decided to enter the hole himself, like if he had a matter to look after. He considered himself a distant relative to the bees anyway, even looked fairly similar to them, he thought nobody would pay any attention to him.

What a miscalculation it was. Barely crossed the doorstep when the apparently empty entrance became full of angry guard bees, and they left no doubt about their intention to chase the intruder away by all means. The wasp wasted no time and hastily spun around in full flight. The guards were on his heels even in the forest for a while. The wasp needed to show a few flying tricks, before the angry followers lost him.

After catching his breath the wasp started to evaluate the situation how he could get closer to the sweet storage. He searched thoroughly the old tree around several times but, was unable to discover any other entrance or crack, which would lead to the desirable storage chamber. Deceiving the guards was beyond his grasp as he had found out before.

Once, the dejected wasp was wandering around the meadow when he spotted a bee flying from flower to flower. At first he didn't think much about it but, on the snapdragon the bee disappeared deep in the flower. The petals completely closed behind her.

Oops! That gave him an idea. He waited until the bee got to the next flower and disappeared again. The wasp jumped there too and held the closed petals tight behind the bee. When the bee wanted to back out she bumped into the closed petals. She tried to force open them in vain. She couldn't understand what was going on when finally realized somebody was holding her captive.

"Is anybody out there?" she asked.

The wasp didn't reply, just waited by holding the petals tight. The bee started to lose her patience. "I ask again. Who is out there? Let me out."

"I'm just a poor relative and I want to talk to you," the wasp said humbly.

"Let me out, otherwise we can't talk."

"I'll let you out if you promise to teach me how to collect nectar and make honey," said the wasp more pretentiously.

"I promise," the bee said hesitantly. "It's not easy because, some abilities are needed to perform such a task."

"Do you promise?" demanded the wasp.

"I do promise," the bee agreed.

When the bee came out of the flower, she astonishingly recognized that her captor was a huge wasp. But, promise was promise and the bee didn't have any choice.

"Collecting nectar is not difficult, I'll show you in a moment," started the bee. "The nectar is at the bottom of the stigma, but not every flower has a nectar and not all the time. You never know and we have to probe it ourselves too," the bee explained and showed it at the same time.

The wasp watched the bee closely and mimicked everything the bee did. He was lucky, the first flower had some nectar. The wasp was delighted, but he expected it to be sweeter.

"You are thinking about honey, which is made in our nest. This is just a nectar, its in a diluted form," the bee educated the wasp.

"How can I get into your nest without being killed by the guards?" worried the wasp.

"The guards recognize you by your smell. I rub our colony scent all over your body and nobody will pay any attention to us."

"Don't trick me, because I will hunt you down and kill you," the wasp threatened the bee.

"I keep my promise. Do you want to see it or not?" the bee replied.

The wasp was more greedy than prudent, so he agreed. Everything went that way the bee had predicted. The behavior of the guard bees didn't change as the bee passed by with his strange companion. The heart of the wasp was pounding so hard like a big drum and he was afraid of being uncovered by that alone. As the wasp eyes became accustomed to the darkness of the bee nest he saw with amazement a huge amount of combs hanging from the ceiling.

"These combs are still under construction for honey storage," explained the bee.

"What's this comb made of?" asked the wasp.

"It's a wax. I think you can produce something like that too, if you rub your abdomen hard enough. Just try it."

"Where's the honey storage?" asked the wasp and started to rub his abdomen as the bee suggested.

"It's way up there. We need to move higher, but you see a little flake of wax has appeared between your abdomen rings," the bee said.

The wasp was delighted to see the results and was convinced he was about to unlock the mystery of honey making.

"I feel a strange smell," cried a bee suddenly near them. "Me, too. Me, too," sounded all around.

"You rubbed off the colony scent of your body, you fool," whispered the bee. "Run, run!" the bee urged the wasp, "otherwise nobody can save you."

The wasp hastily dropped to the floor and turned to the entrance. But, the guard bees jumped on him right away and the wasp couldn't escape without being stung a few times.

Ever since, although the wasps lick nectar from the flowers and can make comb

for their brood, have never been able to learn how to make honey. Since they can't make honey and store food, the wasps perish every winter.

Honeybee

Charles Brunn

I neither sow nor reap
But am part of every harvest.
I am not prey nor predator
Yet there is no feast without me.
Every sop and every sup,
Every dinner and every diner,
Each spirit lifted, and each uplifted spirit;
They are all a part of me. (I am part of all)
I am honeybee, sylph to the sun,
Keeper of the seasons, rhyme to
Nature's reasons. I fly forth at
First light; I set down by starry night.
I am honeybee, winged jewel of the air.
My father has no father. My sisters
Are legion, my mother is my sister.
The sum of me is greater than the parts;
Sans parts, I am nothing.
I am honeybee, sweet is my store.
Each scent, each taste, each touch
Of flower, field, tree and grass
Within my honeyed combs lie
For dearth, for winter, for man
To partake, perchance ask.

A Sickness in the Countryside

Sam Bassett

It was a slow, hot day in the summer of 1886. I checked my watch and sighed when my timepiece read 2:30. My wet handkerchief offered me no relief from the muggy hotness of late. I sucked on a piece of candy in order to distract myself with the sweetness. An ungentlemanly growl escaped my lips when I heard the downstairs door open and then close with a *snap-rap* followed by the groan of the 17 stairs that led to 221 B. I buttoned my collar and tried to act the part of a gentleman despite the heat. Hopefully Holmes would be pleased with having a case, though I was not excited about wandering London in the heat. Perhaps he would not take the case and I would be able to stay melting next to the window.

When I answered the door, a young lady stood before me. In my greeting, I offered her a seat whilst I fetched the great detective. There were few things that would keep Sherlock Holmes inside, but one was the summer heat.

He gave a scathing scowl when I entered. Holmes' black moods were not something to be taken lightly, though neither were my moods. I stomped over, handed him his pipe then grabbed his left Indian slipper which hid his tobacco. Yes, it may have been too hot to smoke but it would awaken his mind.

The woman, Lorelei Clogsmith, explained that her family had recently had a downturn of fortunes. Some of their farmland had been bought by manufacturers. She and the family were substantially paid but, months after, the family had been feeling ill. Chronic nausea, headaches, and kidney problems plagued them. There were no symptoms that would lead to death, but the entire family continued to feel the same illness, though the animals were not ill. A search of the manufacturer's land showed nothing could have obviously gone into the family's food. Also, it was some kilometers downhill from the Clogsmith's farm which prevented runoff into the water supply.

"I think we would do well to take a trip to the countryside. Don't you, Watson?" asked Holmes.

I would take any chance to leave the stuffy city. A trip to a more open space may improve both of our moods. We collected the address from Miss Clogsmith and promised to meet with her in a few days' time.

A train and carriage took us to a bed and breakfast, though Holmes assured me that we would not stay long. After unpacking, we took the small trip to the Clogsmith's farm. It was a quaint little thing with only a few cattle, a garden of nettles, a beehive, and a field. Miss Clogsmith showed us the lands assuring us that we could go as we pleased. Holmes sniffed about for any sort of clues and when he was satisfied with what he had seen, heard, smelled, and, yes, even tasted, we made way for the factory at the bottom of a sloped hill.

For a factory, it was rather clean with only a few spills of liquids on the floors and pieces of debris that were normal for a manufacturing business. Upon inspection, I was surprised at what was being made; the clunky automobiles that were taking place of carriages around the country. I had no affinity for them. They made the air smell worse than the horses and were quite dangerous. Holmes loved the monstrosities because their working parts interested him. If 221B were larger, I am certain he would try to make his own Model SH as I assume he would call it.

The owner of the factory, Mister Edgar Warrick, was a stout man. He seemed to be as wide as he was tall. His mustache was stained by the oil rather than being naturally that dark. He wheezed as he walked and he explained to us that this was just a small start-up automobile factory used to test new designs rather than mass productions.

Given the times, I wondered if this small automobile factory might be a cover for something more nefarious. *The London Times* had made reference to some unrest on the continent of Europe, particularly with the Germans. I had my theories. A strange manufacturing company shows up in the countryside…sickness in the family neighbouring the factory…a war soon to explode out of Europe…

"Ah!"

Holmes' cry shocked me from my theories. He was lying on the dirt; nose only centimeters away from a liquid on the ground with shoes digging into the weeds. It looked to be a liquid that was stained with rainbow reflection. I knew, even with my knowledge on automobiles, that this was not water, oil, or gasoline.

"That's antifreeze. It's new from France. It helps keep the engine from overheating," Warrick explained with great gusto.

"Are there ill effects from consuming this?" I asked while swatting away a bee.

"Well…yes, but there is no way that that has happened. I keep a close watch. The neighbours would not get into mischief here. Should someone consume the antifreeze, they would die within a day or so and no one in these parts has died so quickly."

"But the antifreeze would taste sweet should one consume it correct?" Holmes asked over his shoulder.

Warrick was surprised by my dear friend's words. It was a look that I had grown accustomed to when Holmes spouted out truths from seemingly nothing. I bid Warrick farewell and followed Holmes back up the hill towards the Clogsmiths' residence. Warrick gathered his things to follow, but we were not waiting.

My companion strode past Miss Clogsmith towards the kitchen. For a moment, I wondered if he had become peckish, but that was very unlike him. While working a case, he rarely ate. I watched as Holmes stabbed his finger into the jar of honey and then sucked the condiment off of his forefinger. At once I chided him, though it was all on deaf ears.

"Miss Clogsmith is this honey from your hive?" he asked.

"Yes Mister Holmes. Why would we buy honey elsewhere?" she asked with a confounded look which I reflected as well. "I know that it is ours by the color. Our honey is lighter than the honey at the market."

"And do you know why this is Miss Clogsmith?"

Holmes had always had an interest in bees. It took months to have him agree that a beehive on the rooftop was not a good option and that Ms. Hudson would never agree to it. Even still, I was often suspicious of any honey he brought home.

"The flowers. The flowers around here are sweeter to them I think," she replied.

"Wrong. They are getting another food source. The factory was killing your father and yourself slowly."

"Now see here!" I turned to see Warrick. "I have done no such thing!"

"Accidental I assure everyone. Watson, you remember what I said about the antifreeze."

"That it was sweet?" I offered.

"The bees ate antifreeze as if it were nectar from a flower due to the sweetness!" Miss Clogsmith exclaimed. "But how do you prove it?"

"Perhaps I may borrow some of your candies that you packed in your suitcase my dear Watson?"

I blushed and went to my bags to fetch what he requested. I had grown particularly fond of wax candies which traveled very easily in my pockets. One only had to cut or bite open the wax and a colorful flavored liquid came out that was sweet. My favourites were the red coloured candies. Holmes took the candies and walked outside. I watched as he cut open all of the candies and poured the sweet syrup into the puddle of antifreeze.

"Should I be correct, the hive will create honey that is tinted red. If I am wrong, nothing will change in the hive. She can tell us if I am correct by letter."

Holmes had finished the case in his mind. I was still unsure of the "proof", but it would take some time. I had never heard of bees creating honey of a different colour, but my friend assured me that they would make honey coloured by what it consumed. As we were walking, I saw a large bee buzz about. It did not seem possible to me that Holmes' prediction could be right, but upon further thought, no one else knew as much about bees as my friend.

Weeks went by and the small mystery went into the fogs of memory. It was not until I was eating buttered toast and tea one morning that there was a delivery from the Clogsmith's farm. In his dressing gown and bedclothes, Holmes sat in his chair with the package on his lap. A grin formed on his thin face as he opened the box.

"Thank you, Mister Holmes. Lorelei Clogsmith" was written on a note tied to the lid of a jar of wax-candy-red honey.

Learning From the Bees
Tulip Chowdhury

Growing up in a remote village of Bangladesh in the early seventies was a challenge. Our only access to water was a pond nearly a mile away. We had no electricity or gas and cooked on earthen stoves with dry leaves and twigs as fuel. Life did not have these comforts but we had a vast resource of natural beauty around us. Our village, sitting on a hill, was surrounded by mountains and marsh lands. Along with long stretches of green forests we had plenty of birds, butterflies, bees and other small animals. All the flowers on the green land around us looked like a colorful carpet of nature. Our cottage was surrounded by trees and wildflowers of all kinds. The fresh mountain air was heavenly. It never failed to give me a second awakening to the day every morning as I stepped out into the yard.

When the wind blew, I would breathe deeply to inhale that glorious crisp air and fill my lungs. As the cloying scent of the flowers reached my nose, my ears would pick up the buzzing of the bees and the sweet tweet of the birds. It would be an orchestra of the birds and insects. We had three small steps on our front door that lead down to the yard. All around the front and backyard flowers, butterflies and bees assembled with a festive air. I would settle down on one step and watch the bees as they flew up and down on the flowers.

The trees around me had birds fluttering up and down. While the birds sang with their high and low pitched songs, the bees buzzed endlessly. Their buzzing would be high or low, depending on how near or far they were moving. I could sense strength in the bees' determined movements and their steady buzzing. Most of them were honey bees collecting nectar from the flowers. Occasionally a bumble bee would fly with its loud buzz and send the smaller ones scurrying around. I felt sorry for the smaller ones. The 'right of might' was very much a part of nature even where those insects were concerned. When the bees sat on flowers with soft stems, the flowers swayed as if welcoming the bees. Together, they portrayed the pleasure of giving to life and its needs. The flowers seemed to be so happy when the bees sucked out their nectar as they spiritedly danced with the wind. They radiated more brilliance in their colors after the bees sat on them. I had no doubt that the flowers felt fulfilled with their giving. At school we were taught to be like flowers, to learn to give from what we have, to be whole hearted with generosity. The flowers had life and so did the bees. But the flowers did not turn away when the insects sought food from them. Watching the giving and receiving of life in the beds of flowers around me, I would feel my own heart open up. When the wind blew, and the sunlight surrounded us all, we were a huge part of creations. All those were lessons to me, to be kind to life; to live and let live.

I knew that the bees had one queen and rest are workers. As I watched, I would wonder who was who. But and then when the bumble bees came, they were like giants from another land, ready to conquer the flower fields. The bigger bees symbolized the presence of force in life. The little girl within me understood that in the game of survival one had to fight for one's existence. But there was no menace in the gentle play of force as the bees shared the fields of nectars. The honey bees gave away to the bigger ones, and moved back when the bumble bees flew away. The Earth was home to us all and life was mostly about sharing what nature offered us.

Bees were busy folks just like the little ants running along in lines beneath the trees over which the bees fluttered around. On a mango tree nearby by there was a beehive. Countless bees buzzed and fluttered all around it. It was growing bigger day by day. I knew there was honey stored inside it. It was amazing how those tiny insects were carrying mini drops of nectar and storing it all in that growing hive. Sitting for long hours, my eyes never tired of watching them fly on and off that hive. I was sure all the bees did not take honey from our yard, they went to other distant places to gather their honey. The colony of bees held the lesson of unity and hard work for me. There was so much harmony in nature and how those insects survived through diligence. I also knew that my grandfather would have the hive down and sell the honey for a high price. I would be sad to see the hive go down for I had watched how hard the bees worked for their home. But they did their job and let go to the game of life. There was the lesson of each of us doing his or her own part in life and do it to the best of our abilities.

Bee stings could be bad but those honey bees never stung me. It was, as if they knew I was not there to hurt them. At times one of them would buzz around over my head, giving me some messages in their own language. Bees pollinated the flowers and were the means of continuation of life. I found the bees over my head as symbols of love. Love and reproduction often came together. I was tiptoeing onto the teenage years and whispers of romance floated to me as I watched the bees go through the unique plans of the Creator. In our folk songs came the lines, "Bees come to flowers looking for honey, and I come to you love like the bees…"

People find life's meanings in various ways. My childhood and the bees left milestones in my growing years. I shudder to think how it would be, if bees were no longer to be found, if the pollination stopped or if honey ceased to exist. Honey was the secret remedy for many health problems of humans. Without bees, there would be a big void in our life indeed. What a big part of our life is filled by the little insects. No life is small, each unique its own way.

Applied Beekeeping

Sharing a Cell

Dr. Christian W. TwoRivers

I am a G.E.D. and Functional Literacy teacher at the Arizona State Prison in Buckeye, Arizona. I am also an inmate who has been incarcerated since 1987. I often tell folks that Jesus saved me, but teaching made me the man I am today. Recently, bees have become a part of my life, a passion that I'm sure will play a part of my future.

This past summer, my supervisor was replaced. As with any change of administration, it's a good time to address desired changes to the classroom. I wanted to acquire some better tables and chairs from the warehouse and, as it often is with any state position, it required more than a little cajoling. Since a trip to the warehouse requires going through several secure fences, the acquisition of the furniture turned into a grand production. The students treated it as a field trip of sorts, complete with a dusty old warehouse replete with trash and treasure. Inside were stacks of office furniture, some broken, some not, but everything untouched for an extended period of time. The students had brought along bottles of cleanser and rolls of paper towels and set about picking out some decent chairs and tables.

A few minutes into the adventure came a shout from the back of the building. "BEES!" Several students broke for the front door—fun time was over. For me, the buzzing was like a siren's call as I walked towards the hum rather than away. There was a winding path through mountains of dusty and disheveled furniture which I followed some twenty yards to the back of the warehouse. There lay an upturned recliner, the lining torn away from the boxy bottom. Behind the veil was a tiny world, teeming with life. Mesmerized, I approached as if in a trance. There was a perfect dust-free square on the floor, obviously where the chair had rested for quite some time. The tin wall had been peeled back so as to reveal a window to the outside world, access for the bees to come and go from their comfy home. I gently replaced the chair back into the dusty outline on the floor and backed slowly away, all without a single sting.

Several students had been stung. Nothing more serious than the fright of actual pain. We procured the furniture we came for and left the warehouse behind. The little adventure was great fodder for tall tales and raucous laughter back on the cell block. But I had been stung with a metaphorical sting, one that drew my mind back to that alien world behind the veil. This was something to write home about.

After hearing of my adventure, my dear mother purchased a subscription to *American Bee Journal* for me. I began reading about "the hygroscopicity of honey", how "one out of every three mouthfulls of food on the planet owes thanks to pollinators", and just "how lucrative beekeeping can be". After reading two issues, I was scared to death of varroa, small hive beetles, and wax moths! I manipulated passage back to the warehouse which, in this prison setting, is like returning to North Korea or something. I simply needed to know about the health of this feral colony.

I learned a lot on my second trip. First off, I'm not the bee whisperer I thought I was, and I'm not immune to bee stings! More interestingly, I found no varroa, beetles or moths. This led me to believe that my bees were domesticated, treated and were probably from the nearby crop fields that surrounded the prison. They were escapees! Having fled the exploitation of some large corporate enslaver, they had set up shop in a place where they would "bee" understood. They live in cells, we live in cells. They like sweets, I like sweets. Of course, having more than one active queen has a totally different meaning for them than it does on the prison yard. But all in all, beekeeping is very conducive to prison life. For me, watching the little guys go back and forth to the free world, right through "no man's land" and the razorwire fences… it's like a little taste of freedom. And in the privacy of the dusty old warehouse, nobody can hear me talking to them and asking what its like out there as they pollinate the local soybean, hay and cotton crops.

Over the past few months, I've returned to the hive half a dozen times. I harvested two nice chunks of comb early on and was amazed at how quickly repairs were made. I'm pretty sure that my bees are *Apis mellifera ligustica*, the Italian bee, so for a nostalgic comfort I sing "Ave Maria" ever so softly as I tend them.

I traded a pouch of loose tobacco for some confectioners sugar from the mess hall, mixed solution for them and poured it into a plastic trash can lid, placed a piece of old bubble wrap in it and created a feeder. It was a thrill to see these creatures use something I designed. I have *ABJ* to thank for the education, and the ideas.

My last trip to the hive was before Thanksgiving. Around the holidays, when people go on vacations, it's hard to gain access to areas that require an inmate to be supervised. I'm hoping I did well in preparing the hive to go dormant. Here in Arizona, there really isn't much of a winter. In the coming weeks, I'll try to get back to the hive. In the mean time, I have made good use of what I've learned.

In the classroom, I teach all of the core subjects, but science now has a new flare. I have incorporated bee science into numerous lesson plans and have even participated in small business classes given by the local community college. Beekeeping has relatively low start-up costs, and it's great therapy in the endeavor. For me, finding this hive has been a life changing experience.

Commercial Keeping

William Blomstedt

The maple honey flow was on, Charlie said. His friend Kevin from across the county told him the supers were filling up. We had planned on shaking some packages today to get a head start on the pickup at the end of the week, but we hadn't been to maple country this year, and, with a flow, the hives would be busting. Eight of us stood under the mimosa tree listening to Charlie's morning spiel. Last week it had been nine, but Raymond let the vodka get a hold of him a bit early on Friday—I could smell it on his breath from across the truck—and gave Charlie too much of his mind. Now Ray's probably in his trailer on the Bonn property, in the same state but worse. He needed to go, but in a way I felt sorry for him. 20-odd years of hauling heavy boxes and getting stung without seeing much advance. This commercial beekeeping could wear you down.

The heavy Georgia morning weighed on the entire group. I wasn't sure about the rest of the crew, but last night Jeff and I parked our truck at around 8 PM. We had spent the day clearing a few yards which hadn't been used for years, whacking through the chest-high grass to find broken pallets, piles of sun-melted frames and every now and then a living beehive. Of course these would be marked with a big red X: American foulbrood spore-breeders. By the end of the day, the Freightliner's bed was filled with junk and we could only park it and head to the bunkhouse. I took off my boots, ate a sandwich and watched about five minutes of TV before getting sucked under. Some hours later I was able to take off my jeans and stumble into bed but, after a blue minute, the alarm hit and I was up, pulling on the same jeans, getting my foot stuck in the hole in the knee and making it rip a bit more. My hands still buzzed from the weedwhacking and the rest of my body was some combination of sore and tired, except for my lower back which just plain ached. I needed to see the doctor again, but there probably wasn't time this week. Nor the next. I would have wait until package season was over.

The commute to the honey house was just a few hundred feet across the field. That hundred meters was awful short sometimes. The other guys got in their trucks, turned off their phones and drove away at the end of the day. But the seasonal workers, like Jeff and I, stayed in the bunkhouse on the property and were fair game for extra chores at any hour. "Like cell phones—working nights and weekends free." as Jeff liked to say.

When Charlie had finished his spiel, the group dispersed: some off to catch queens, the rest to check the hives across county, and I to unload the Freightliner and bring pallets of comb behind them. The engine switched on without fuss, thankfully, and I drove past the bunkhouse and down to the clearing with the burning pit. After untying the ropes, I put on a veil, climbed into the Bobcat and bumped it off the trailer. First I lifted the pallet with three red-x'ed beehives off the truck and placed them on the charred-out center of the pit. I wadded up a chunk of newspaper but my sticky lighter wouldn't give off more than a spark. I had to search through the entire truck before finding an ancient box of matches. The first few match-heads crumbled, but finally one lit and I set the burning paper on the pallet's empty spot. Soon, a few dark, old combs got the blaze blowing and an armload of broken boxes turned it into a bonfire. The sides of the doomed hives began to blacken and bees spilled out the cracks. Instead of flying away, they just hung onto their home until the fire licked them to a crisp. All those little lives. I unloaded the remaining pallets of junk and began to toss more on. Some of it could probably be fixed, but no one in this operation would do it. Soon the fire was as tall as the Bobcat and I watched it for a few minutes before reloading and bouncing back to the shop.

On the other side of the property, I parked on the bare dirt in front of the comb warehouse. The overhead door was off its tracks again and it took all my strength to push it up. Inside was a hellish place: a 30x30x30 foot windowless cube filled with comb stacked on pallets. Each pallet could hold 60 supers, if they were stacked ten-high, and the pallets could be stacked on top of each other—grimy skyscrapers reaching into the cobweb-darkness. A chemical smell overran the wax, but the paradichlorobenzene we dumped in every month still somehow didn't kill the black widows, brown recluse and a whole horrorshow of other creepy crawlies that populated the warehouse's cracks and comb. It was not a place to be without gloves.

I brought the Bobcat inside and surveyed my task. The space was small for this machine, and the cluttered, collected junk on the floor made it even smaller. Two pallets of deep comb would be easy to get, but another was sitting on the third story and at an awkward angle. The fourth was currently unseen and would require some Tetris-like maneuvering to find. I pushed the forks of the skid-steer inside the first pallet, lifted it, and gingerly backed out.

While putting the first two pallets on the Freightliner, another was revealed in the second row. I moved one pallet on the ground, shifted another over a few feet, and I was able to sneak it out with a few spare inches on either side. After putting it on the truck, I returned into the warehouse for the final pallet. To access this one, I slid an entire three-story tower a few feet to the left, holding my breath as the giant moved in one piece. Then, peeking through the cage above me, I was able to guide the forks into the final pallet and lift it gently. Unlike a forklift, which picks up pallets vertically, the Bobcat's arms lift at an angle, so the driver needs to manipulate the tilt of the forks simultaneously to keep the pallet, and untethered towers of supers, level. This took me many hours and a few dropped pallets to master, but this time it went smoothly. I inched back then lowered the forks until the pallet rested in front of me. Deep breath. I had it. But then, backing out of the warehouse, I felt the back wheels drop and a wave of momentum forward.

"Oh, fuck." I said and froze, but next I heard BANG... BANGBANG. The goddamn lip between the pavement and the dirt was a two or three inch drop and deepens with every traverse. This was the final test and it hooked me. I cut the engine, lifted the safety bar and climbed out to survey the damage. Nearly four columns had fallen off the pallet. They laid in a messy, sharp heap of wood and wax. Somehow, and this always seemed to happen, every fallen comb had jumped out of its box. There was no easy way to put them back in. One at a time, nine per box. How many boxes? But before I could begin, my phone rang.

"Yes?"

Charlie. "Where are you? I've been calling the last twenty minutes."

"The Bobcat was loud. Just tying down now."

"OK, never mind the comb. That son-of-a-bitch Kevin is crazy, there's no flow here. In fact there's not a drop of honey left in these hives and we need to feed them. So take those deeps off the truck and start filling cans. We're gonna need, oh, 10 boxes here. And that probably means Wildwood, Gravel Pit and Longly all need to be fed too. So grab Juanita and you two fill everything we've got. The boys will be back to pick up the first load."

Sigh. "OK. Bye." I hung up and looked at the chaos in front of me. Comb back in the boxes. Hundreds of cans of sugar syrup to fill. The only bees I worked with today I had burnt to death. All these things needed to be done to be a beekeeper, but I sure didn't feel like one. An itch appeared in the back of my throat, one that cried for a beer. Maybe I'd bring a six pack over to Raymond's place tonight, to see how he was doing.

To Course A Bee

William E. Jones

Winging free across the meadows,
Charging sun before it shadows,
Signaturing sole and soul in morning dew.
Taking flight with thoughts high soaring,
Fleeing boorish mores so boring,
Finding solace in the mystery I pursue.
Taking bearings on my mountain,
Where there flows a golden fountain,
With coordinates encrypted by the bees.
Hidden somewhere in this pasture,
Reigns a queen with court and castle,
This enigma harbors clues in sumac trees.
In July with sumac blooming,
Nectar-hungry bees are humming,
This ambrosia lust will lay their kingdom bare.
When they sup in heightened rapture,
Homeward lines of flight I'll capture,
They will lead me to their honey laden lair.
Through my covert observation,
I perceive their destination,
Having broken figure-eighted, wag tail codes.
Nectar laden paths I follow,
Down through meadows cross the hollow,
And in one hour stand before their oak abode.
This old tree I mark with pleasure,
Proving ownership to treasure,
My invasion ends devoid of axe or saw.
With my initial on your tree,
You may reign as sovereign bees,
Protected by bee hunter honored law.

In Service to the Queen

Joan Riise

In the ancient land of milk and honey it was known far and wide that Sakinah of Nazareth, trusted midwife and healer to the families of Galilee, was also their keeper of bees. In days long past, Sakinah's mother Saya had called wild swarms to be her own, nurturing them in hand-crafted earthen cylinders, a method she'd learned from her Egyptian mother a generation earlier.

Through her mother's and grandmother's stories, gentle wisdom and lifetimes of experience, Sakinah had learned both healing skills and how to care for her beloved bees. She valued her bees not only for their liquid gold but for the healing qualities their honey, pollen, propolis and venom provided. Through her ministry, many an elder had experienced welcome relief from the crippling diseases associated with aging. Pregnancy and afterbirth recovery were enhanced with the richness of her harvest and in those days, babies thrived on both mother's milk and raw honey.

And it came to pass that in that time there went out a decree from Caesar Augustus that the whole world should be enrolled.[1]

Sakinah had been visiting with Mary and Joseph for some time as her pregnancy advanced. Mary was young and strong, carrying her child easily, but she was so close to delivering when the village decided to join a caravan to make the trip to Bethlehem together. Sakinah would travel with them on the 100 mile journey taking the longer passage between the towns for safety from bands of thieves known to victimize travelers along the shorter route. Anticipating any possibility, she packed supplies thoughtfully with salves of honey, beeswax and oil of almonds, poultices of propolis, healing herbs and clay, lavender and hyssop to sweeten the birthing bed and several goatskins of her legendary honey wine. The trip would be arduous but Sakinah would do her best to keep Mary comfortable and the baby safe while easing Joseph's concerns.

Making the journey with caring neighbors and friends could have been enjoyable had not the urgency of Mary's condition weighed heavily upon everyone. The nights under open skies at least afforded respite and the pleasant exchange of stories around the fire. Mothers and grandmothers eagerly regaled the couple with tales of their own miracles of birth.

Close to ten days they journeyed, Mary traveling on the back of a donkey to keep her from exhaustion. Sakinah was always close at hand though she traveled by foot. She understood only too well how important it was to maintain not only Mary's strength but her spirit as well. Much work lay ahead.

And it came to pass when at last they arrived, no room could be found for them at the inn.[2]

Urgently scouting what might be available, Joseph eventually found safe haven for them in a barn. The kindness of the owner afforded them a bed of clean straw, light, water and nourishment. Relieved, Sakinah offered one of her goatskins of mead in gratitude for their lodging bringing the hint of a smile to Joseph's weary face. The warmth given off by the animals and their gentle lowing calmed the family as Sakinah made preparations for the imminent birth.

Over the next hours, a sacred silence descended upon the small circle of family as Mary labored through the night. The birth was not difficult for one so young and strong but it was her first and as such proceeded slowly. Sakinah encouraged Mary while keeping Joseph close. With a sensitivity born of experience attending labors beyond number, she became increasingly aware that this birth was like no other. The mother though young, expressed no fear. She uttered no sound. Her breathing was even and deep. Though she responded easily to Sakinah's every signal and suggestion, she seemed to be attuned internally as if in prayer or meditation. A glow emanated from her body, its light illuminating their surroundings. Peace prevailed even through the hardest passage.

And she brought forth her firstborn son and wrapped him up in swaddling clothes and laid him in a manger.[3]

The miracle foretold through the ages had been accomplished. Sakinah cleaned the beautiful, perfect boy child handing him first to Joseph.

Overjoyed, he spoke from his heart in silent communion with his son. When Mary was made comfortable, Sakinah handed the babe named Jesus to his mother who took him eagerly to her breast. The sound of wings and heavenly voices filled the air!

And behold, an angel of the Lord stood by them and the brightness of God shone round about them. And suddenly there was with the angel a multitude of the heavenly host, praising God and saying: Glory to God in the highest; and on earth peace to men of good will.[4]

Her work accomplished for the time being, Sakinah retreated to rest herself amongst the warmth of the animals, allowing the holy family opportunity to bond. Sleep would not come easily that night as there was round about them the divine presence of angels singing. Shepherds arrived who on their night watch in the hills had heard the angelic host announcing the long awaited birth. They came bearing gifts of new lambs for him who was called King. Within a few days' time, the magnificent wise men arrived from the east who had followed the star. They came with gifts of gold, frankincense and myrrh.

Sakinah remained with the family caring for Mary and Jesus, witnessing their joy and wonder until eight days were accomplished and the time of circumcision was at hand. So many intimacies and gifts had been shared among them during those days. Sakinah was humbled at the privilege she had been allowed. As it was soon her time to rejoin the caravan and return to her home in Nazareth, she pondered what gift she could possibly give to this family and the baby Jesus who she would love for all her life. As she was packing her things for the journey home she finally understood what it was she had to give. Ready now to depart, she approached once more the Holy Family and withdrew from her robe the jar of honey and a tiny vial she'd nestled in her basket in what seemed like a lifetime ago.

"Mary, Joseph, and you dear, sweet baby Jesus, you have touched my life at the source of my spirit and in ways I do not yet understand. I am honored and grateful! You have received gifts from kings and shepherds and in leaving you, I too want to share my wealth. Please accept this jar of honey and with it this precious vial of Royal Jelly. The bees and this work of birthing are my life. Especially with the bees, I've always felt blessed to work for the Queen but you chose me, your humble servant, to help you deliver a King. My wealth is not measured in riches but through you my soul has been enriched beyond measure. My hope is that I live long enough to follow his growth through the coming years in Nazareth, our mutual home. Please accept my gift knowing that these small treasures will always flow from my Queen and her bees to you and your Son. May your angels guard and keep you safe on your journey to Jerusalem. Farewell for now, sweet, holy family. I eagerly await the time of your return."

The End and the Beginning…

Author's note: though it is not written historically, religious scholars consider it highly unlikely that Joseph and Mary would have made the journey from Nazareth to Bethlehem alone, a distance of 80 to 100 miles, possibly 10 days in duration. With what Joseph and Mary knew of the impending birth, traveling in a caravan with a midwife in attendance would ensure the greatest degree of safety for mother and child.

References

[1]St. Luke, chapter 1, verse 1

[2]St. Luke, chapter 2, verse 6

[3]St. Luke, chapter 2, verse 7

[4]St. Luke, chapter 2, verse 9, 13, 14

Experts

Deborah Darling

The electrician knows
how to avoid hot current
that kills on contact
and I, the sting,
having studied bees
as he has the energy
that brings us light
or death.
Skillfully,
he touches circuits
that might sizzle,
stop his heart,
And I handle, unafraid,
the home of small friends.

SSSSURPRISE!
Phill Remick

In 1978, I worked for a large commercial outfit near Fresno, CA. It was a super wet year; the kind beekeepers dream of and dwell on. Old timers declared that, during these rare years, the bees "would make honey off the fence posts".

We had moved the majority of 6,500 colonies into various spring locations and two of us were headed with a couple hundred strong colonies toward some superb black sage and buckwheat locations near Salinas, California.

That night we would load other colonies just down the road and ferry them home.

Bees were kept primarily on massive ranches, often in very rugged, hilly and undesirable terrain; finding a semi-level spot to place them was a miracle! Often, hives were unloaded by hand because the area was too steep to stabilize. The hive loader attached to the truck could easily swing out of control, dragging its operator along for the unwanted ride.

The rainfall totals that spring in California were off the charts and it seemed most of our five bee trucks were stuck or spinning in the mud the majority of the time. On numerous occasions, we had to have the loads of bees, usually 120 colonies per truck, towed in and out of the water-saturated and flower-laden areas by tractor. What a mucky mess!

In the past, I had witnessed many wild critters behind these private gates, including a rare mountain lion. We traversed in slow motion toward the bee location and I was lucky enough to complete my mission with no problems, thanks to a tractor escort! Meanwhile, my partner, Carl, was to deliver his group two miles south of me in a dry, flat, semi-green area. The balance of the day, we relaxed and that evening we ate some dinner before driving home to the Central Joaquin Valley. We were both planning to pick up and return more hives that night which were required to balance out a very large alfalfa seed pollination contract in Corcoran, California.

As I returned to the dilapidated green wooden gate meeting place with my load, I noted the overwhelmingly dark sky coated with an incredible mantle of stars blazing above. This was the splendor of California 1978.

I waited for Carl. I waited and waited. We had no radios or cell communication which made a situation like this rougher to endure. Where was he? Was he hurt? Was the truck damaged?

I adjusted the seat in the 2 ton truck with its 20 foot flat bed and began to snooze. Then Carl's truck came thundering down the road like the proverbial 'bat out of hell'. Dust flying, he rolled up, bounded from his truck, looking very spooked. Then I learned what had happened.

Our hive loaders were made of two flat, fork-like pieces of steel which slide under the hives while the main frame fit against the hive side with support from a 4-inch square chunk of tire rubber that automatically (spring loaded) clamps down on the top of the hive. We could then press a button on the handle to raise and lower it into position on the truck bed while walking it to the vehicle, using that fold down handle which was about an arm's length, as a guide.

My scared bee buddy said he had slid his loader forks quickly under a hive (it was one of the last ones of the load) and he remembered the top clamp tightening as he began to move the hive which was right at waist high, toward the truck bed. Then he heard something strange.

I looked at him and said, "Ok so what was the problem? Did the loader cable snap or motor burn out?"

He looked me dead in the eye and said," I picked up a rattlesnake by the tail that was under the colony. It was whipping back and forth near my stomach. I really thought I was done for."

I said, "Oh, man, that's terrible, what did you do?" He replied, "Well, I was (expletive) fortunate enough to have the majority of the squirming, writhing, and angry snake to the right of me. I hit the down button and let the hive crash to the ground. I ran to the truck, pulled it forward and tied it down; I figured I'd rather get stung than risk a rattlesnake dangling from my side with fangs firmly embedded."

I stared at him in total shock as he calmly explained, "So, looks like we're one hive short on the load tonight."

Taste of Chicago
Sampling the Law with a Chicago Urban Beekeeper
Jason M. Shimotake

On Wesley's roof, you could see the sun peeking over the Chicago sky-
line in the distance. What was usually so impressive off the highway or
on the back of a ninety-nine cent postcard, from Wesley's roof in the
Ukrainian Village, looked so small you could wipe it away like spilt salt
on the dinner table. For a moment, between being up so early and given
a view of the city that I had never seen, I didn't notice the hum coming
from Wesley's direction. I turned to Wesley and took a step back. He
was still wearing his jean shorts and cut-off tee but had added a giant
helmet with mesh netting. From the neck up, he looked like a spaceman
heading to a funeral. From the neck down, he looked like a scrawny kid
who mowed lawns for a nickel. With deliberate hands, he moved a heavy
brick off the top of the box. With his other hand, he held a metal can.
Smoke wafted from it. Wesley directed the flow into the box. The hum
was enormous for a single moment but slowly died down, the smoke
dampening the noise. I couldn't look as Wesley moved his hand into the
box. After what seemed like ages, he put the lid back on the top. He put
the brick on top again. I sighed in relief.

Wesley took the mask off. He was smiling. He held a small mason jar in
his hands. A brown swatch rested against the inside of the jar. It oozed
brown juice, which started to collect at the bottom of the jar. Wesley
opened the jar and stuck a wooden stick into the jar. It drizzled down the
stick as he handed it to me. I fumbled with it as I brought it to my mouth.

It was a heavy taste, like an earthy wine. Its sweetness was complex and
coated my tongue. It tasted like fresh lake water, like daisies in Wicker
Park, like the lilies from the Garfield Park Conservatory. It tasted like
Chicago and I couldn't help but smile. It was a hell of a sales pitch. And
from Wesley's self-satisfied grin, he knew it, too.

It was business that brought me to the pleasure of Wesley's delicious secret. Wesley Greenwood worked in the Accounting Department and although we never met outside of work for more than the company kickball league—footnoted as the only team in firm history to be completely shut out—he insisted on coming to my going away party. I was a junior attorney at the firm and was leaving to go work for another firm (they had a softball league, which sealed the deal for me).

A few of us, including Wesley, went down to McKinney's Pub, one of the old Irish joints that stuck around like barnacles to the hull of downtown Chicago. At half past five on a Friday, the bar was starting to bulge with patrons. The Bulls game blared from the strategically-placed televisions and hubs of grumbling businessmen hunkered around the bars. After a few drinks, the group thinned out and the well-wishers started to go home. At that point, Wesley bought me one last round. At first I thought he was coming onto me, as he wasted no time in inviting me to his roof as a "consultant" for his new business. Seeing as I was finding myself with a little time on my hands, I thought, "What the hell?"

After our Saturday morning meeting, I agreed to do some research for Wesley. We met at Peat's just down the street from his apartment/apiary. I ordered a larger black coffee. He sat down with a large carafe of tea. He spooned in some honey.

"You brought your own stash?" I asked.

"Oh," Wesley laughed, "no, this isn't mine. This stuff might as well have come out of a squeeze bear."

"A snob, are we?"

"Well, when you've had the best…"

"So I did some research about your specific business in the city of Chicago and I was surprised to see how much was out there, but my first question is how did you get into this in the first place?"

"What do you mean?"

"I mean, it's kind of weird, right? You're in the middle of a city, not on some farm in Michigan or something."

"Well, actually, there's a pretty large beekeepers community in Chicago. A lot of the high rises downtown have hives on top of them. It's kind of a conservation thing. Even the mayor was doing it for a little while. My girlfriend was working for NPR and did a story about them and I kind of tagged along. Funny, I only got off my butt and started building my hive after she left me though. It's weird how things like that happen. Anyway, I started watching some YouTube clips and just started building a kit and next thing I knew I had a queen and it was like I couldn't stop."

"And when was this?"

"Oh, like two years ago. This is my third summer."

"Okay. Well, what do you want to know?"

Wesley sipped his tea as he digested my question. Clearly, there were a lot of questions all shoving to the front of the line. After another moment, he spoke.

"I mean, I guess my thing is I think I want to like, do this as a business. With this one hive, I got like sixty pounds of honey. I gave a lot of it away but, in the end, a lot of it just went bad. So I figured I could sell it to local coffee shops or farmer's markets and stuff. And eventually, I'd like to have like, two or three hives up there. So am I like, okay to do that?"

"Well, there's a lot to consider here. First off, the City of Chicago's silent as to having hives in the first place. But it advises that you check with your neighbors on that. Have you talked to them about your hive?"

"Yeah, the other two units in my building are cool with the hive I have now. Mrs. Rodriguez grows flowers in the backyard so she's all about it."

"But if you're going to have two, you're going to have twice as many bees. If anyone complains, they might deem the hives a nuisance and shut you down pretty fast. In fact, Skokie almost banned hives altogether. Thankfully, they didn't but it shows how sensitive of an issues this can be. So be careful. Also, and this isn't really legal advice, but I read that bees need large amounts of water. You should have a large water source. The article I read said that if you don't, they might seek water from air conditioner units.

"In my opinion, it's about avoiding as much conflict as possible. You should keep accurate records of equipment used, how your bees act and have those records saved so if you have an inspection, you'll have that available. And once you register, be ready for inspection. You could get fined for not being available. Speaking of which, you know you have to register your hive, right?"

"Uhm," Wesley shrugged, "no."

"Well, you do now." I picked up my phone and typed in some text. "And here's where you do it[1]. You could face some stiff penalties if you don't.

"As far as selling, I don't see why not. You probably want to start a corporation or non-profit to protect yourself. And here," I handed him a small packet, "I figured you were interested in selling, so I printed out the Rules & Regulations for Chicago Farmer's Markets. You basically apply just like a street performer, but in my opinion, protect yourself so you don't get stung, pardon the pun."

"Whoa," Wesley said, "that's like, a lot of information."

"Yeah," I nodded, "take it for what it is. Free advice is as valuable as how much you pay for it."

Wesley nodded and nodded. For a moment, he looked like a bobble head doll swaying in the air conditioning. This happens to every attorney on occasion. Information overload. I asked Wesley if he had any other questions, but he just shook his head. At that point, I was about to burst so I excused myself to go to the bathroom. When I came back, Wesley was already gone. But in his place was a small Ball jar of honey. I unscrewed the lid and tasted Chicago.

Disclaimer: This is a work of fiction. In no way is any of this work to be construed as legal advice. All characters appearing in this work are fictitious. Any resemblance to real persons, living or dead, is purely coincidental. But if you're interested, Wesley is single.

Endnotes
[1] http://www.agr.state.il.us/programs/bees/beekeep.pdf

Lessons Learned

Helen Ackermann

The day began like other days for a couple of retired people. My husband said, "Well, it might be a good day to go the bees." I might have had other plans but, as the assistant beekeeper, I knew we should attend to the "kids" as my husband has occasionally called them.

We started out; it is about a ten mile drive into the countryside where we keep our bees. They live on the hobby farm of a friend with opportunities to pollinate apple trees and gather nectar from a variety of garden crops, basswood and goldenrod. It is an idyllic place to have our bees.

The day started out being a sunny one but by the time we arrived we could see storm clouds gathering in the far west. We put on our hats and veils. I normally wear the hat and veil, long gloves and a long sleeved shirt. My husband indicated that we should hurry as the weather seemed to be changing. We were there to take some honey, not always a popular activity with the bees. We began checking the frames and the bees seemed rather ornery. It wasn't long before I noticed that a bee had found an opening in my veil and was crawling in my hair. I moved away from the hive, but bees followed me and soon there were a number of them in my hair. I walked quickly away but they followed and soon were stinging me. I called to my husband who wasn't sure what he could do. I finally removed by hat and veil and tried to swat the bees. He even smoked me, but the damage was already done. I was stung about twenty times on my neck, face, ears and on my head. Needless to say, we went home without completing our task.

I am not allergic to bee stings but it was a very painful experience. I put ice packs on and just waited out the pain. By morning my face was so swollen that I could not see properly and was not able to take our grandchildren to school as I had promised our son. Within twenty-four hours I was feeling much better and the swelling was beginning to recede.

Later that fall, our grandson, asked us to come to his class to give a presentation on beekeeping. We were happy to oblige and it became a wonderful lesson in the importance of following directions. After my husband shared information on bees, showed the class frames, the smoking device and brought them a treat of honey, I told them about my experience. I did not blame the bees but rather reinforced the truth that it is important to follow directions. I was to blame because I was not careful when I donned by hat and veil. I was in a hurry and did not pay attention. I reminded them of the importance of following directions when at home and in school. There is a reason for those directions and disregarding them can lead to experiencing all kinds of problems.

I wondered if I would be afraid of the bees after my encounter with them, but I was not. Bees are quite wonderful but one has to know one's bees, pay attention to the weather and follow directions. Bees teach us many lessons about life and what I have learned is that it is important to don one's hat and veil carefully.

The Mindfulness of Beekeeping

Jennifer Ford

I tend to have problems with focus. My mind jumps around from one subject to another like a hyperactive squirrel. While walking the dog, I am making lesson plans for my eighth grade classes. While trying to correct papers, I am planning on what to have for dinner. My mind is swirling with emails to return, phone calls to make, chores to be done, and projects to be accomplished. Meditation never really worked for me—sitting and trying to empty my mind just felt like one more thing to do.

Then I realized—I already had a way to practice staying more focused right in the yard next to my house—my honeybees. Going out to the beeyard to open my hives is a lesson on remaining completely in the present moment.

As I enter the beeyard, I breathe in the smell of honey and beeswax and hear the hum of the bees. I see a steady stream of bees flying out over the neighboring field, and returning with pollen. As I puff smoke in the front of the hives, I am focused on the behavior of the honeybees. Are they bringing in pollen? Do they seem aggressive at all? Are the guard bees at the front entrance? There is no lesson planning going on now.

I gently remove the outer, then the inner, cover and say hello to the bees. I move slowly, carefully, and deliberately. When I first began beekeeping, one beekeeper told me that when you are in the beeyard opening a hive, you should move like you are underwater. No hurried, staccato movements here.

The bees line up between the frames, and begin to take in the honey that has spilled from the broken comb. I carefully scrape off the burr comb with my hive tool, mindful of bees that may be in the way. I lift my veil and taste a bit of the honey that is leaking from the comb. Dinner? Who cares?

Next, I begin to inspect each frame. Those that are loaded with honey are given a quick glance, but I focus my attention on the brood frames. I see eggs smaller than a grain of rice, larvae curled up in each cell like the letter "c", and capped brood tucked away under beeswax until they are ready to emerge. Pollen is packed into cells in shades of white, yellow, and orange. Uncapped honey glistens. Emails and phone calls are far, far away.

I search, sometimes in vain, for her majesty, the queen. This requires laser-sharp focus—to find the one bee with the elongated abdomen among all the others. Sometimes I spot her, but often I need to be content with just finding her eggs. Sometimes in beekeeping you need to let go of the outcome, and be happy with the process.

Once I am satisfied that I have gathered all of the information I need about that hive, I begin to carefully put it back together. I smoke the tops of the hive bodies and supers to get any stragglers out of the way, and slowly, carefully, return each level of the hive to its rightful position. As the outer cover is returned, I take a breath, gather my tools, and head for home.

I remove my gear, and sit down with a cold drink and my notebook. While it is fresh in my mind, I record my notes about the hive. I reflect on what I observed in the hive today, and make a plan for my next trip out. When I finally glance at a clock, I am usually surprised by how much time has passed. While I am inspecting the hives, everything else, including time, seems to cease to exist.

Keeping bees has helped me to become a more sane, focused individual. Carrying the lessons learned in the beeyard into the rest of my life helps me to focus, to pay attention, and to stay in the moment, no matter what external circumstances may bring. Beekeeping, without a doubt, is my favorite form of meditation.

PASSING THE SMOKER

Only the Bees Know

Greg Leatherman

From a damp acre full of meadowsweet,
we carried cedar boxes to the sourwood groves,
where galax and arbutus bloom,
and up the ridge followed the swarm
from juneberry blooms to giant yellow-poplars.
In spring, we brought back morels,
sassafras and spikenard, but by July,
the mid-day heat drove us to the peach orchard,
where again we set the hives.
In late summer, we hunted quartz crystals
in the ruts of an abandoned logging spur
that carved through matted thickets,
to the edge of a trickling hollow.
Where mushrooms swell with shade
and the ground squishes,
we sat upon the stump of a fallen yew
whose giant roots sprawled across a bed of brown needles
before diving into the loam.
Our shoelaces covered in burrs,
we laughed at the woodpecker tapping.
We sliced mush-melon
and crushed peppermint leaves for lunch.
And when we'd had our fill
and the light leaking through the treetops turned,
Lily began gathering dead branches
from a black locust tree
growing out of a long, mossy mound.
"Only the bees know who is buried here," she said.

Brave New Worlds
The Bees and Me
Linda Butler

When I was I child I dreamed of becoming an astronaut. I wanted to visit strange new worlds and meet exotic creatures. Bees were some of the first aliens I encountered when I stepped out of my cardboard rocket ship into my back yard. Bees were both frightening and fascinating. Their black and yellow stripes and their buzzing erratic flight made me think that they were little extraterrestrials, buzzing around in spaceships. And those tiny creatures carried an awesome advanced weapon—a ray-gun-like stinger that could zero in on an unsuspecting human and inflict intense pain and possibly even death! Bees might be alien, but they were also amazing!

One of the most incredible things about bees was they could fly around to flowers and trees and magically produce a delicious liquid gold—honey! My favorite lunch was a peanut butter and honey sandwich. When I was about five years old, a couple of the neighbor boys convinced me we needed to become avid hunters of the delightful and dreadful *Apis mellifera*. Armed with plastic pails and shovels, we set out to capture bees and harvest our own honey. One sunny spring Saturday, we traipsed through hillsides of colorful California Iceplant. The plants were in full bloom, their bright pink, yellow, and magenta blooms attracted hundreds of bees. As avid *Apis* hunters, our task was to smash the bee with the shovel and scoop it into the pail. Later, we would dissect the bees and remove all that delicious honey!

A morning's worth of work yielded only a thin scattering of bees at the bottom of our pails. As young as we were, we realized this was not a profitable way to procure honey. The dissections were dismissed and we returned, disappointed, to our homes.

A few years later, when I was twelve, my family and I went camping at Yosemite National Park. While on a nature walk, the Forest Ranger commented on a fallen and decaying log. "The knots of an old pine log are wonderful to burn in a campfire, giving off more heat and a lovely piney scent," he said. Then and there, I knew we "needed" an armful of knots to enhance our evening campfire. I found a stick and eagerly poked, prodded, and worked several knots loose.

Suddenly, angry buzzing filled the air as bees streamed from the log began to attack me. Over a dozen fiery darts of pain shot through my hands, arms, and legs. I swatted and I cried. When a bee entered my ear, I panicked. Not just an angry buzz, this bee sounded like a rocket engine, boring and barreling its way through my head. If the bees didn't kill me then and there, a sting inside my ear would certainly render me deaf for life.

I don't know if whacking myself on the side of my head or the calming hands of the ranger and my parents got the bee out of my ear, but eventually I calmed down and limped back to our campsite. I was given a dose of antihistamine then I flopped down on a sleeping bag to rest. I survived, but for the rest of the day I felt achy all over and had a dozen and a half painful, itchy spots on my arms and legs. I realized bees deserved a certain amount of caution and distant respect.

Life went on and bees and aliens faded into the past as I dug into my present life of college, marriage, children, and career. Most of my jobs involved working indoors in schools and libraries but I found a refreshing, part-time job working as an assistant tree pruner to a neighborhood arborist. He had kept bees while growing up and living in Canada, so one of our customers asked for his help on managing her new hives. To tree assistant, I added bee assistant. The first time we opened and examined a hive, my childhood fascination with bees was instantly rekindled. To be holding a brood frame with thousands of creeping, crawling, and working bees was amazing. To see an entire society in boxes of less than a cubic foot each enthralled me. We cared for the bees throughout the summer and held a "honey harvest party" with the help of the customer and several of her friends.

When my arborist friend and neighbor, Roy, decided to keep a couple of hives on his property, I became his assistant here, too. It was a great partnership. There was a regular beeline of bees traveling from his hives to my pond. We harvested several gallons of honey from the hives and felt great satisfaction and camaraderie in caring for these fascinating and delightful creatures. Roy was the capable teacher, and I, the eager student. We attended beekeeping classes. We read numerous books. We discussed and searched for ways to improve the lives of our apian charges.

When Roy suddenly died last year from a heart attack, I became the main provider and keeper of the bees. It was sad but healing to follow the old world custom to kneel beside each hive of bees and gently tell them that their beloved keeper had died, but that I would be there, caring for them in his place.

When I was young, I had dreamed of being an astronaut. Space ships, in the form of little buzzing bees or huge roaring Saturn V rockets, faded away as I married, had five children, was widowed, and raised them alone. But somehow, as I don my white bee suit, cover my head with the hat and veil and my hands with the gloves and enter the strange square worlds of my beehives, I feel somewhat like an astronaut, meeting and mingling with these alien apian societies.

Blessing of the Bees

Hannah Valentine

I am not a quick decision maker. Most quick decisions have led me into danger, such as running a red light during a driving test or chasing a horse down a busy highway—on foot. As such, I have learned from my mistakes.

The day was not particularly pleasant. A wet, windy autumn had tricked all the Iowan trees and fields into a deep, brown sleep. Air leaching in through the van windows smelled sweet with decay. We pulled off the country road after Justin's pickup truck. My dad, my brother Josiah, and I stepped out onto the squishy ground, and my dad put on the traditional black robe Orthodox Christian priests wear before grabbing his prayer book. In the bed of Justin's truck lay various tools of his hobby: a manure bucket, a smoker, and a white suit which resembled those of astronauts. He handed me a few items to carry and, seeing my smile, asked, "You ready?" I nodded excitedly and we followed him into a nature preserve.

A low-lying structure, like a hay manger, hummed before us. Justin called it a "top-bar hive." Several tiny blurs of black and yellow zipped through the air around us. After lighting his smoker with manure and newspaper, Justin pumped a few gentle puffs at the entrance. He explained how the smoke distracted the bees; in fear of a fire which might destroy their honey, they would gorge themselves and care less about who peeked into the hive.

When Justin lifted up the roof, followed by the first bar, my heartbeat accelerated. I saw a golden pendant of wax and honey. Crawling across its surface, the hive's thousands of honeybees hurried to salvage their food. Justin asked if I wanted a closer look. My hands quivered as I zipped the astronaut suit on, finishing with a netted hat on my head. Then I stepped forward. The hive was another world, a world held together by paperthin honeycomb and its fuzzy engineers. A perfectly beautiful community.

But this community was sick. Having not seen the queen during his past few visits, Justin had asked my father to come out and say a blessing for the hive. Of all the insects in the world, only the honeybee receives a special blessing in the Orthodox Christian prayer book. Their importance to the church can not easily be overlooked; much of the intoxicating aroma one can smell during a service comes from the traditional burning of beeswax candles.

My dad opened his book but remained a few paces away, heeding Justin's warning that the bees might mistake him for a black bear in his robe. "O God," my dad prayed aloud, "who knows how to work benefits through human labor and irrational living things, You instructed us in Your loving-kindness to employ the fruits and works of the bees for our needs. Now humbly we beseech Your majesty: Be pleased to bless the bees and increase them for the profit of the human race, preserving them and making them abundant. Let everyone hoping in Your majesty and Your boundless compassions, and laboring in the care of these living things, be counted worthy to receive abundant fruits of their labors and to be filled with heavenly blessings in Christ Jesus our Lord, to whom is due glory, honor and worship unto ages of ages. Amen."

Following the prayers, Josiah handed our dad a bowl and a long-fiber brush. Josiah poured out a bottle of blessed water into the bowl. Then, taking a few brave steps forward, our dad dipped the brush in the water and sprinkled the hive, "In the Name of the Father, and of the Son, and of the Holy Spirit." My dad nodded to Justin with a smile; we were done.

A simple blessing.

Justin closed the hive, and I returned the suit to him. On our way back to our cars, Justin thanked us for our time and expressed his mixed feelings about the hive's condition. "It's really frustrating, you know? There's nothing I can do to intervene and make sure that they survive the winter. But it's kinda like life, some things are just out of our control." Then he smiled at me. "Maybe, if this hive doesn't make it, you would be interested in using my equipment to start beekeeping."

We said our goodbyes and wound our way back down the country road into town. The whole ordeal had only taken a half an hour. A simple blessing. The fate of the bees rested in God's hands. Yet I made a rather impulsive decision on the ride back. Or perhaps my heart had been set the moment Justin showed me the first honeycomb. I knew next to nothing about beekeeping. My bank account was nothing to boast about. But I knew that I would start a hive of my own. I would learn how to form mutual trust with the bees. Then, I would be free to marvel at their incredible feats of complexity and simplicity for years to come.

Adeline's Honey Harvest

Rachael Gaude

'Tis early in the morn when Father awakes me to help tend to the bees. Today we harvest the summer's honey. I quickly dress in my usual bee-keeping attire—my gardening dress, layers of woolen socks, and a thick coat—to protect me from the bees' stings. Once outside, Father hands me old leather gloves, stained from past beekeeping adventures, and a straw hat draped with coarse cloth to shield my face.[1] Father is dressed similarly, but in breeches of course. He directs me to bring a wicker basket with the tools needed for harvesting while he carries a barrel to hold the freshly harvested honeycombs.

The path to our beehives is lined with crepe myrtle trees, whose fragile pink flowers blow into my hair with the early breeze. Our hives sit along the far edge of the garden. Virginia's terrain, as well as our garden, offers a variety of flora for the bees. Most flowering plants in the garden I know by name—tulips, primroses, peaches, and apricot blossoms in the spring, followed by snapdragon, prickly lantana, and cleome in the summer.[2] Wildflowers bloom almost year long. The differing plants that bees in the colonies draw nectar from are as diverse as the china in Mother's cabinet.

From a distance, our hives look like large tree stumps lined against the garden's fence. They are stumps, to be frank. In the springtime, Father explored the nearby forests to collect bee colonies that had made their wild nests in hollowed trees.[3] He took that bee-filled portion of the tree home and fitted the top with a wooden slate to protect the contents from rain. These tree hives, called "gums",[4] are more functional than the earlier English methods used when bees were first brought to America, or so I have been told.

Father once explained to me that honey bees were not in the New World before the first settlers came. The Virginia Company in London likely shipped the pioneering bees alongside other livestock to the settlers sometime in the winter of 1621.[5] Those bees were kept in skeps as the English had kept theirs.[6] A skep was a like a large woven basket—about a forearm wide and two heads high—and made of straw or wild grasses. Swarming, which occurs when a hive becomes so large that it must separate, has helped the honey bees spread across the colonies.

Once near the hives, I can hear lively buzzing inside the gums. From my basket, Father takes a small sack filled with pieces of sulfur. I sit on the wet grass while he sprinkles the contents of the sack at the entrances of all but one hive. Using flint and steel, he lights the sulfur pieces. Sulfur creates a putrid odor and it kills the bees inside;1 we will not be stung when we collect their honey. "You forgot one," I tell him.

"Not so," he corrects, fanning a little sulfur pile with his hand. "The hive left untouched will grow until it swarms early next spring. We will catch those swarms to replace the ones we are losing now."1 I nod so he knows I understand.

Father sits next to me. He and I linger a long while in the morning sun, waiting for the gums to grow silent. Suddenly, Father smiles. "Think of all the honey and beeswax we'll have, Adeline."

"Aye," I say. "Think of Mother's honey cakes!"

"Cakes?" he chuckles. "Of all the goods we acquire from the bees, you think of cakes?" Father kids with me because he knows I am well aware of honey's extraordinary uses. Besides using the honey for tasting and cooking, Mother uses it as a preservative, making cured hams and jellied fruits.[3] Local breweries make mead by fermenting honey in water.[7] Doctors employ honey in their treatments, for it cures burns, soothes upset stomachs, reduces coughing, and eases sore throats.[8]

To colonists, the leftover honeycomb is as valuable as the honey itself. Beeswax candles are the difference between an early retirement to bed and a prolonged evening. In addition to candles, the wax is rubbed on leather[3] to make water roll off the surface. It can be run over a thread for smoother sewing,[3] dress wounds, and relieve pained joints.[9] Distinguished ladies like the ones in the governor's mansion even use beeswax in their wig creams or something odd like that.[10]

Eventually, we no longer hear the bees' humming and we begin to extract the honeycomb. Using a knife, Father cuts away heavy portions of the comb from inside the gum. He hands each to me, and honey drips on my shoes while I sweep away the dead bees with my glove. I place the intricate honeycombs into the barrel and we proceed to the next hives.

When all the combs are harvested, Father carts the full barrel into the kitchen which is near the garden. He lifts the barrel onto the worktable with the help of our slave, Rose. "The bees have done their work, now it's time to do ours," Rose says, gathering bowls for hand-crushing[1] the combs and cloths to strain the honey.

The three of us roll our sleeves and spend several hours pressing the honeycombs to drain the golden liquid from its waxy home. The wax is placed on coarse cloths which will be bound tightly and hung on a rafter so that the excess honey can drip from it.

As many hours of crushing creep by, I never grow weary of the sultry scent of beeswax that fills the hot kitchen. Keeping honey bees is wearisome work—from beginning early in the morn, to dressing thickly; from gathering the tools, to waiting for the gums to become still; from cutting out the combs, to—at last—crushing the honey. I smile, press my finger into a honeycomb, and finally steal a taste of the long-awaited honey. It adds an unmistakable sweetness to my life in colonial America.

Endnotes

[1]Oertel, Everett. "History of Beekeeping in the United States." *Beesource.* Revised 1980. Web. Accessed January 2, 2014. www.bcesource.com

[2]Young, Joanne and Taylor Biggs Lewis Jr. *Washington's Mount Vernon.* New York: Holt, Rhinehart and Winston, 1973. Book.

[3]Pilling, Ron and unknown editor. "Bee Skep: Making a Bee Skep for Your Garden." *Colonial Sense.* Copyright 2014. Web. Accessed January 2, 2014. www.colonialsense.com

[4]Marchese, C. Marina. *Honeybee: Lessons from an Accidental Beekeeper.* New York: Black Dog and Leventhal Publishers, 2009. Book.

[5]Kellar, Brenda. "Honey Bees Across America." Oregon State Beekeepers Association. Website copyright 1999-2012. Web. Accessed February 3, 2014. www.orsba.org

[6]National Beekeeping Centre Wales. "History of Beekeeping." *About Bees.* Unknown year. Web. Accessed January 27, 2014. www.beeswales.co.uk

[7]Sutherland, Roger. "A Taste of Beekeeping History." *Ypsilanti Gleanings.* Published in Ypsilanti Gleanings, Winter 2012. 2012. Web. Accessed January 28, 2014. www.ypsigleanings.aadl.org

[8]Molan, Peter C. "Honey for the Treatment of Infections." *Internet Archive.* 2007. Web. Accessed February 4, 2014. www.web.archive.org

[9]Yelsukova, Irina. "Beeswax Use." *Martelcom.* 2000. Web. Accessed February 4, 2014. www.mari.su

[10]Westminister City Archives. "Unpalatable Pumeatom." *The Cookbook of Unknown Ladies.* 2013. Web. Accessed February 6, 2014. www.lost-cookbook.wordpress.com

Bibliography

Carlson, Laurie. "Review of Tammy Horn's 'Bees in America: How the Honey Bee Shaped a Nation.'" *H-Environment, H-Net Reviews.* 2007. Web. Accessed January 2, 2014. www.h-net.org

Crowder, Les and Heather Harrell. *Top-Bar Beekeeping: Organic Practices for Honeybee Health.* Vermont: Chelsea Green Publishing, 2012. Book.

Dadant & Sons. *Beekeeping: Questions and Answers.* First Edition. Illinois: Dadant & Sons Inc, 1978. Book.

Flottum, Kim. *The Backyard Beekeeper's Honey Handbook.* Massachusetts: Quayside Publishing Group, 2009. Book.

Krebs, Bill. "Bees in the Colonies." *Colonial Williamsburg, Past and Present Podcast.* 2009. Web. Accessed January 10, 2014. www.podcast.history.org

Pellet, Frank C. Edited by Dadant & Sons. *American Honey Plants.* Fifth Edition. Illinois: Dadant & Sons Inc, 1976. Book.

Pottoroff, Laura Pickett. "Some Pesticides Permitted in Organic Gardening." *Denver County Extension Master Gardener.* 2010. Web. Accessed January 28, 2014. www.colostate.edu

The Waggle Dance
Laurie Wallmark

Wiggle, waggle, wiggle.
    Field bees follow me.
    Come and find the flowers
    out beyond that tree.

Wiggle, waggle, wiggle.
    Blossoms straight ahead.
    Yellow, pink, and purple,
    but we can't see red!

Wiggle, waggle, wiggle.
    Here's the nectar site.
    Golden drops of treasure.
    Honeybee's delight.

Robin's Bees
(A *True Story* of Urban Beekeeping)
Therese Calegari

"Mommy!" seven year old Lisa called excitedly to Therese Calegari.

"What is it?"

"You've gotta come see what Anna found—in our driveway!" she shouted. "Robin's got bees!"

Therese went to see. Lisa and her four year old sister, Anna, were both pointing at a big crack in the roof of an alcove window attached to the living room next door. Bees were quietly flying in and out.

"Robin!" Therese called down the driveway next door. An old man with a pipe in his mouth holding a chair and a piece of sandpaper looked up from an open garage.

"What the hell do you want?" he said, taking out the pipe. Therese grinned. "What?" he said. "I'm in the middle of doin' somethin.'"

"Come see your infestation," said Therese, smiling.

Robin came. "Holy smokes," he said, peeling wood glue from his fingers. "It's a hive."

"What are you going to do?" asked Therese.

"I don't know." Robin looked again, then shrugged. "Nothin' I guess." Therese broke into a laugh.

"Robin!" she said. "Aren't you worried about your house? Your hair is covered in sawdust, by the way."

"Well, you stopped me in the middle of sandin,'" said Robin. "And don't you go laughin' about my answer. You're the one who's always complainin' about people wrecking nature in the world. Well *there's* nature. It's just an alcove roof. It's hardly attached. They're not *in* my house. As far as I'm concerned, the bees can have it."

Therese laughed more. "Robin, you are such a weirdo—you're going to leave *bees* in your roof?"

"Think what you like. I'm an old man and I've told you my opinion. But you probably want 'em out, do ya? Since they're in your driveway."

"I'm not sure."

A lot of neighbors gave Therese advice. "Make him call an exterminator," said one. "Bees don't belong here."

"They're too dangerous!" warned another. "Especially in your driveway. What if the children got stung?"

"What if they're those horrible Africanized bees?" said a third. "You could be killed, you know!"

A bee buzzed slowly past. It didn't seem very Africanized.

Therese hadn't grown up in the city, and she hadn't grown up in the suburbs. She'd grown up on a farm in Australia. A lot of bees in Australia didn't even have stings, but plenty did. Farmers loved them. You could easily live with them if you left them alone. Her dad had kept a hive on the ground outside their garage when she was little. She'd been taught not to go too close. She'd never been stung. Therese thought about the driveway. Apart from when she and the children gardened, nobody used it. On the side next to Robin's house was a long boundary wall: seven feet high at the back and waist high in the middle below where the bees were. Taller than Lisa or Anna.

"Let's try leaving them," Therese told Robin. "I don't use the driveway, and I'll teach the girls to be careful. But if your house falls down, don't blame me."

"It won't. And if you change your mind," said Robin, "I'll call someone, okay?"

"Okay." So the neighbors tut-tutted and gossiped about the eccentric old man and his impulsive friend, and the bees stayed.

Two years passed. The bees were left to do their work and nobody was stung. The only real change Therese noticed was that the orange trees in her backyard gave more fruit. Then one day, a big storm came to Pasadena. The wind blew hard and the rain pelted down. The alcove roof with the bees began to sway. A piece cracked and fell. More fell, then more, until the whole thing collapsed and slid into the gap between Robin's house and the boundary wall. Bee-coated pieces of roof were scattered everywhere.

"Now what?" said Therese.

"Well, we can't leave 'em there," said Robin. "But it'd be a shame to lose 'em."

"You like building stuff," said Therese. "Maybe you could build them a hive. Then we could get honey."

Robin laughed. "Honey!" he said. "What do I look like, a beekeeper?"

"You could be," said Therese. "You'd make a great beekeeper. Look at how happy they've been in your roof. This is obviously a good spot for bees. Come on, think of the honey you could give me!" She grinned.

"Ha!" said Robin, but he was smiling too. "Well, I guess I could look into it. But where would we keep them? I'm not climbin' onto my roof to collect honey. Either I put a hive on my roof and ignore it, or if you really want *some* of the honey, I'd have to put it somewhere lower down, like our boundary wall—maybe at the back. On my side, of course."

"Sounds good to me," said Therese. Now the comments were even more incredulous.

"I can't believe you're letting him do it," said someone, "it was bad enough when they were in that roof."

"There's no way I'd let anyone keep bees near children," said someone else.

"I don't let my kids run on the road," said Therese, "and I won't let them climb a seven foot wall to visit a hive on the other side. What's the difference? If anything, it's further away than when they were in the roof. Nobody got hurt then. I will teach the girls to be careful and I trust them to be sensible." The neighbors shook their heads but Therese stayed firm.

For the next couple of weeks, Robin talked a lot to the people at the Los Angeles Urban Beekeeping Society. He bought new wood and sawed and hammered and painted in his garage workshop. The end result was a tall white hive. There were wooden frames that slotted into the top like slices of bread. Each one was covered with special wax paper. The paper was where the honey would grow. At the bottom was a box where the piece of hive with the queen would be placed.

"She might not stay," said Robin. "She might not like our hive. But let's hope she does. And there's another thing I have to tell you."

"What?" said Therese.

"Apparently there were two hives in that roof. So if the first queen doesn't stay, we'll try moving the other one in a few days."

From inside their house, Lisa, Anna and one of their friends watched as Robin and a professional beekeeper climbed over the boundary wall to where the bees had fallen. The two of them were dressed in white and wore special hats with veils. Robin carried what looked like an old oil can. He squirted smoke from the can into the broken roof to make the bees sleepy as they went. The beekeeper reached down and started sorting through the pieces. She held up a piece. Bees were going crazy on it. This was where the queen was. Robin squirted more and more smoke. The beekeeper took the piece and put it in the bottom of Robin's hive. Then she closed the hive and took it to Robin's side of the boundary wall to attach it. She came back and picked up another chunk of roof and broke it open. Rich golden honey drizzled out onto her yellow rubber glove. A lone bee flew onto it and ate a little honey while the children watched. That night everyone chewed on honeycomb.

The queen stayed in Robin's hive. The other queen was donated to a budding Los Angeles Urban Beekeeper.

After six months, the hive had grown a lot. Robin added a box to the middle. He also collected some honey. He gave Therese a jar. He explained that the honey changed color depending on what was flowering when you took it. You couldn't take it at all in the winter because the bees needed it for food. The first jar was pale gold, the color of honey from orange blossoms.

The next year the hive swarmed as a new queen left to make a new hive. The swarm gathered on a tree nearby, then left. The following year Robin built another hive, and the swarm moved to it instead. Soon there were three hives on the wall.

Lisa and Anna were sensible and left the bees alone. "See," Therese told a neighbor, "children can live with bees."

Eventually Robin registered his hives with the City of Pasadena. Now they were protected by law.

The Calegaris and Robin grew to love living with beautiful, wild bees and, with time, the neighbors around them did too. Everyone's orange trees were always heavy with fruit, local gardens were full of flowers, and Robin shared his honey with the whole block. And everyone knew it was all because of Robin's bees.

The Beginning
Katharine Atwood

Out back behind our old gray house, two soft white Langstroth hives lived proudly through my childhood. They hummed with buzzing life for years, their colonies nursing green into our cucumbers and pinks into our cosmos, zipping past our dandelion-filled hands on their way to press their bodies to our roses, buzzing red into them too. As a child, I thought little about them. Their magic was taken for granted. It was like sunshine, like moonlight, like star-light, wind or rain. It simply was.

At seven I began to notice them, curious, at first, of the sound that filled the air out by the boxes I knew they lived inside. When the tending to of children and work had begun to fill the days too full, my father started leaving their wooden frames of honey untouched through winters. They made it, we'll let them enjoy it this year, he'd say, and I was largely none the wiser. The process of spinning liquid from their comb was not one a little brain remembered or yet understood. But when my heart and eyes and ears found themselves all drawn to the edge of the woods out there beside them, he quietly took notice. For you, he said one nearly summer day soon after and handed me a small white canvas suit. For the bees.

I met the bees that year, that summer, and we fell in love. The smoker was my job, my task, and my quiet mind filled to near overflowing joy that first approaching moment with it lit. Stopping unsure before I came too near, he gestured for me to come close, and with the cover lifted off, I saw, like nothing I had know was real, a billion bristling moving bits of black and gold.

A puff, he said, and motioned here to show me how to move my hands like we had practiced by the barn. Slow and careful I brought my fists together and watched the gray breathe out and fall onto them. They slowed, a softness falling also down onto them, and my mouth hung wide and open as we watched it land.

Now here, he said, and lifting up the puffer lay one thick brush into my hand, and pointing to the hive swept his own in one soft motion over the top. Like this, he said, and motioned once again. I felt my eyes grow big, reached up and over and pressed the bristles down soft into the bees. Some moved, some flew. The air was thick with sound and quiet movement. From safe inside my suit, I looked out in awe through netting as they gathered on my arms and hands, moving all about us in some strange determined peace.

There are two jars of honey left from all we extracted that last Summer, still dark and sweet and safe inside of glass.

The bees died the following winter.

And too easily, the living of modern lives overtook our time again, so the getting of a new swarm was pushed from Next Spring to Next Spring, until I was grown and moving out and on my way. Why don't you get more bees? I asked in recent years, talking to my father from the distant world of college. There's an empty house out there in need of lodgers. Where, I asked him, does one find a swarm? The money, he'd say, the time, the energy. There was always *something* that would not let it happen.

The flowers miss their bees, I'd tell him. The planet needs them too. For years we kept this going, him agreeing. But it was never right, never time.

I dipped down into one of those last jars this past Christmas. Cracked the lid with '98 scrawled upon it open. He saw and stopped me in surprise and seeming horror. That's precious! He cried out. It's all that's left from what they made, it's precious, we must save it. I dropped a spoonful, thick and gold, into my tea and stirred it, smiling. So make more, I said, and held it up for sipping. Get bees, I said, and shook my head and drank. He paused and went to speak but didn't.

Last week, he called, the phone line full of waiting joy and coming declaration. The colony, he said and smiled through the phone. We pick it up this April, and I think, he said, you're going to need a bigger suit.

CONTRIBUTORS

A., J.P.	53	Feeley, MFC	138
Ackermann, Helen	12, 185	Forbes, Peta-Ann	8
Ames, Randy	128	Ford, Jennifer	114, 187
Atwood, Katharine	207	Foster, Laquavia	14
Bassett, Sam	157	Gaude, Rachael	197
Blakely, Catherine	23		
Blomstedt, William	133, 168	Haen, Angeline	52, 58
Botto, Zosia	30	Hance, Benjamin	4
Brunn, Charles	156	Harris, Inga	113
Bryant, Karen J.	117	Hawkins, William	10
Butler, Linda	191	Herdan, Sarah	84
		Hoeft, Rebekah	28
Calegari, Therese	202		
Carrier, Roman	142	Ives, Gary	124
Channell, Rachel	74		
Chowdhury, Tulip	161	Jones, William E.	172
Christie, Elizabeth	65		
Citrino, Anna	45	Kuebler, K.E.	98
Cox, Nathan	36		
Crainshaw, Jill	54	Leatherman, Greg	190
Csincsa, Tibor	149, 152	Link, Jordan	39
Darling, Deborah	92, 177	Mattison, Liza	75
De La Garza, Lela Marie	144	Miller, Abigail	147
Dickens, Andrea	111, 137	Montgomery, Carrie	33
Eastick, Erica	106	Nester, Robbi	57
		Newday, Amy	102

Pendergraft, Bill   29

Pichard, Lauren   56

Prybyla, Alixandra
    Nicole   80

Purvis, Michelle   19

Remick, Phill   178

Resides, Jane   17

Riise, Joan   173

Shimotake, Jason M.   181

Skelley, Billie Holladay   48

Stout, Dan   68

Sullivan, Payton   121

Talamantes, Daniel   87, 103

Thompson, Mary
    Langer   44

Trout, Stefanie Brook   93

TwoRivers,
    Dr. Christian W.   165

Valentine, Hannah   194

Wachtenheim, Josh   132

Wallmark, Laurie   201

Walsh, Tarah   72

Wilke, Vicki   35, 42

Wood, Tabitha Peyton   60

CPSIA information can be obtained at www.ICGtesting.com
Printed in the USA
LVOW04s0712240615

443511LV00004B/8/P